Alginate Fibers and Wound Dressings

Alginate Fibers and Wound Dressings

Seaweed Derived Natural Therapy

Yimin Qin

WILEY-VCH

Author

Prof Yimin Qin
Jiaxing University
No. 899 Guangqiong Road
Jiaxing
China 314001

Cover Image: Courtesy of Yimin Qin

Library of Congress Card No.: applied for

British Library Cataloguing-in-Publication Data
A catalogue record for this book is available from the British Library.

Bibliographic information published by the Deutsche Nationalbibliothek
The Deutsche Nationalbibliothek lists this publication in the Deutsche Nationalbibliografie; detailed bibliographic data are available on the Internet at <http://dnb.d-nb.de>.

Print ISBN: 978-3-527-35329-3
ePDF ISBN: 978-3-527-84518-7
ePub ISBN: 978-3-527-84519-4
oBook ISBN: 978-3-527-84520-0

Typesetting Straive, Chennai, India

Contents

Author Biography

Dr. Yimin Qin studied for his PhD at the University of Leeds between 1986 and 1990. After spending three years at Heriot-Watt University working on his postdoctoral project, he became the R&D manager at Advanced Medical Solutions Plc in Cheshire, UK, where he led a team of scientists and developed a number of high performance fibers and wound dressings from alginate, chitosan, and other natural polymers. He then went to study for an MBA at Manchester Business School and, after graduation, took up the position of Fibers Product Manager at SSL International, working on advanced silver containing antimicrobial alginate fibers and wound dressings. In 2002, Dr. Qin went back to China and taught at Jiaxing University in Zhejiang Province. In 2015, he was appointed as the director of the State Key Laboratory of Bioactive Seaweed Substances at Qingdao Brightmoon Seaweed Group, where his main research interests focused on the extraction, purification, modification, and applications of alginate and other novel bioactive seaweed substances.

Preface

Alginate is a natural polysaccharide extracted from brown seaweeds. In the medical textile industry, alginate fibers have been used for the production of high-performance wound dressings due to their high absorption properties and good biocompatibility. In recent years, as more and more efforts are made to use alginate fibers in conventional textiles and industrial textiles, many types of functional alginate fibers have been developed by using ion exchange, polymer blending, and other novel processing technologies. These fibers combine the properties of alginate and various types of additives, making them useful in a large variety of specialized applications.

Technically, calcium alginate fibers can be readily made by first dissolving sodium alginate powder in water and then extruding the solution through a spinneret into an aqueous calcium chloride bath. In addition to this type of conventional alginate fiber, functional alginate fibers can be made with a number of novel processing methods. Because it is a polymeric acid, alginate fibers can be used as carriers to deliver zinc, copper, silver, and other bioactive metal ions for wound care and other novel applications. In addition, since both the dope preparation and coagulation processes are carried out in aqueous solutions at a neutral pH, many bioactive materials, such as drug and enzyme, can be combined into the alginate fibers without losing their bioactivities.

Alginate fibers have many unique properties that are highly useful for functional textile materials and medical textile products. For example, they possess ion-exchange and gel forming properties when in contact with body fluid, in addition to other excellent performance characteristics such as hemostatic, antimicrobial, wound healing promotion, skin whitening, and many other unique bioactivities. These novel properties are highly applicable to medical textile materials such as functional wound dressings, functional face mask materials, hygiene products for women and children, and incontinence products for adults.

Historically, the most successful application of alginate fibers has been functional wound dressings made from alginate fibers through nonwoven processing. These high performance dressings have been widely applied to a wide variety of chronic exuding wounds, where the high absorption is achieved via strong hydrophilic gel formation, which limits wound secretions and minimizes bacterial contamination. In addition, alginate wound dressings can help maintain a physiologically moist microenvironment that promotes healing and the formation of granulation tissue. Upon wound healing and dressing change, these novel wound dressings can be rinsed away with saline irrigation, so removal of the dressing does not interfere with healing granulation tissue. This makes dressing changes virtually painless, a performance highly beneficial to patients with acute or chronic wounds.

This book offers a general introduction to the sources of alginate and the production methods for alginate fibers and wound dressings, in addition to the novel properties and applications of these functional materials in wound management. Taking into consideration the latest results of clinical research conducted around the world, this book summarizes the unique properties of alginate wound dressings, including their "gel blocking" properties and the ability to promote wound healing, facilitate hemostasis, reduce pain, suppress bacteria growth, and lower treatment costs in the treatment of a wide range of wounds including leg ulcers, burn wounds, pressure sores, surgical wounds, and many other types of wounds with high levels of exudate.

Because alginate fibers and wound dressings cover a large field with a diversified range of specialist knowledge, it is inevitable that this book will not be able to offer precise explanations in all areas, and the author appreciates critical feedback in such cases.

30th June 2023

Yimin Qin
Jiaxing University

1

The Extraction of Alginate from Brown Seaweeds

1.1 Introduction

Alginic acid is an anionic polysaccharide distributed widely in the cell walls of brown seaweeds, where it exists in the cell walls and extracellular matrix in the form of a mixed salt of sodium, calcium, magnesium, strontium, and barium alginate. The British chemist E C C Stanford first described the extraction of alginic acid from brown seaweed in a patent dated 12 January 1881 [28]. In the following years, Stanford carried out the initial studies on the chemical nature of alginic acid, which he named "algin." [29] Due to the protein components in the seaweed extract, Stanford initially believed that alginic acid contained nitrogen.

Brown seaweeds are distributed in many parts of the world, and following Stanford's initial work, many other scientists around the world made further studies on this novel biomaterial. In 1926, Atsuki and Tomoda [2] and Schmidt and Vocke [27] reported that uronic acid was a constituent of alginic acid. Shortly after these two studies, other scientists found D-mannuronic acid in the hydrolysate of alginate [3, 17, 19–21].

The chemical nature of alginate was further clarified in 1955, when Fischer and Dorfel found that in addition to D-mannuronic acid, the hydrolysates of alginic acid also contained L-guluronic acid [6]. This finding is important to illustrate the nature of alginic acid as a copolymer composed of two types of monomers, i.e. D-mannuronic acid (M) and L-guluronic acid (G). Figure 1.1 shows the chemical structure of these two monomers.

It is now widely known that as a natural copolymer, the proportions of D-mannuronic acid and L-guluronic acid vary widely for alginate extracted from different types of brown seaweeds, resulting in variations of the physical properties of alginate-based materials. As a polymeric acid, alginic acid can form salt with various types of metal ions to form alginate salt, with alginate being a term commonly used as a general description for the various types of alginic acid based salts, as well as all the derivatives of alginic acid and alginic acid itself. These seaweed-derived polymeric materials have thickening, gelling, emulsifying, film, and fiber-forming properties that are widely utilized in many diversified industries.

Alginate Fibers and Wound Dressings: Seaweed Derived Natural Therapy, First Edition. Yimin Qin.

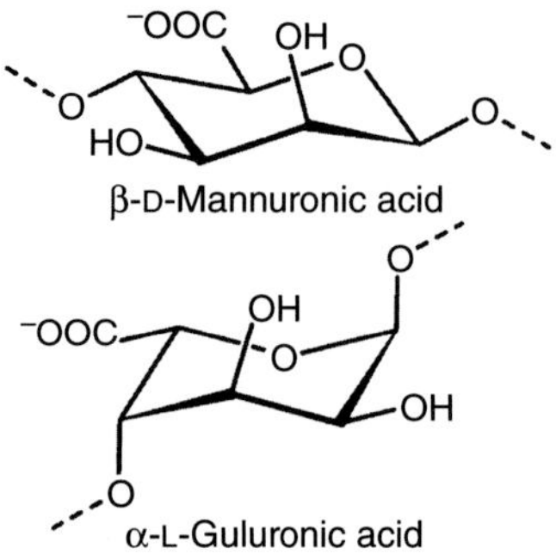

Figure 1.1 The chemical structures of β-D-mannuronic acid and α-L-guluronic acid.

1.2 Global Distribution of Brown Seaweeds

Although alginic acid is present in several types of bacteria such as *Azotobacter vinelandii* and many species of *Pseudomonas* [7] and alginate can be successfully extracted from the biomass of the soil bacterium *A. vinelandii* ATCC 9046 cultivated on crude glycerol as an alternative carbon source [10], up to the present time, all commercially available alginates have been extracted from brown seaweeds, mainly *Laminaria hyperborea, Macrocystis pyrifera, Laminaria digitata, Ascophyllum nodosum, Saccharina japonica, Ecklonia maxima, Lessonia nigrescens, and Durvillaea antarctica*. The chemical composition of alginate extracted from different types of seaweeds varies according to seasonal and growth conditions, as well as within different parts of the plant [1, 13]. This variability in composition shows the biological role of alginate in seaweeds whose mechanical properties can be regulated partly by the variations in the M/G composition of alginate. For example, the brown seaweed *L. hyperborea* grows in very exposed coastal areas, where high mechanical rigidity is required in the stipe and holdfast, whereas high flexibility is needed in the leaves that float on streaming water. The stipe and holdfast contain alginate with a high content of L-guluronic acid, which is responsible for high gel strength, while the leaves contain alginate with a high content of D-mannuronic acid, which is related to softness and flexibility.

Figure 1.2 shows the distribution of the main species of wild brown seaweeds around the world. Globally, brown seaweeds, or Phaeophyceae, are a large group of multicellular algae that play an important role in the marine environment, both as a source of marine vegetables and for the habitats they form, which provide a natural environment for other marine organisms. For example, the brown seaweed *M. pyrifera* may reach 60 m in length and form prominent underwater forests. In the Sargasso Sea, the brown seaweed *Sargassum* creates unique habitats in a vast area of tropical water. In many parts of East Asia, the brown seaweed *S. japonica* is widely used as food for human consumption. Overall, there are about 1500–2000 species of brown seaweed in the world [11].

Figure 1.3 shows the main types of brown seaweeds used for alginate production. In 2015, the worldwide annual industrial production of alginate was estimated to be

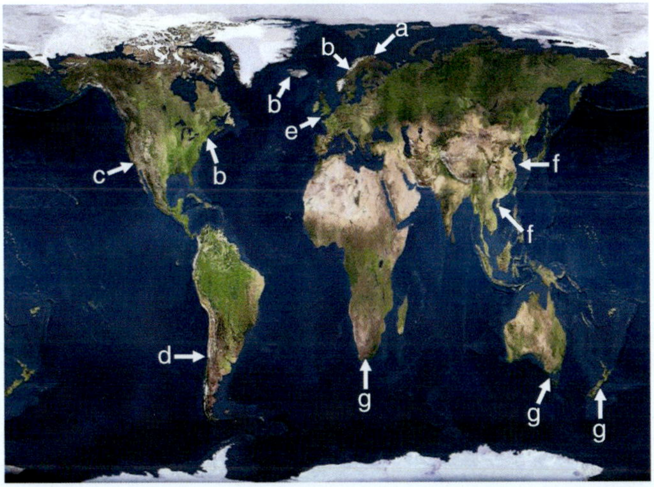

Figure 1.2 Distribution of wild brown seaweeds around the world, (a) *Laminaria hyperborea*; (b) *Ascophyllum nodosum*; (c) *Macrocystis pyrifera*; (d) *Lessonia nigrescens*; (e) *Laminaria digitata*; (f) *Saccharina japonica*; (g) *Ecklonia maxima*. Source: Adapted with permission from fig. 1.4, Qin [26].

Figure 1.3 Main types of brown seaweeds used for alginate production. Source: Adapted with permission from fig. 1.2, Qin [25].

55 500 metric tons, utilizing 236 820 tons of dry brown seaweeds [26], which represents a small percentage of the biosynthesized material in naturally occurring wild brown seaweeds. It is estimated that the total quantity of wild seaweed stock on the Norwegian Sea coast is around 50–60 million tons, with 7 million tons washed to the shoreline annually. During the algae bloom in 2019, 20 million metric tons of brown seaweed, known as the Great Atlantic *Sargassum* Belt, stretched almost 9000 km. These naturally available seaweed biomasses are complemented by the cultivated seaweeds, and it is reasonable to assume that alginate is an unlimited and renewable resource even for a steadily growing industry.

The properties of alginate vary from one species to another, and the choice of which seaweeds to harvest and cultivate is based on both the availability of particular species and the properties of the alginate that they contain. For example, alginate extracted from *S. japonica* is not suitable for the production of food gel due to its high content of mannuronic acid. At present, the main commercial sources of wild brown seaweeds are species of *Ascophyllum, Durvillaea, Ecklonia, Laminaria, Lessonia, Macrocystis, Sargassum,* and *Turbinaria*. Of these, the most important are *Macrocystis, Laminaria,* and *Ascophyllum*.

Macrocystis pyrifera grows best in calm, deep waters with temperatures of 15 °C or less. It is sensitive to water temperature and does not withstand a rise above 20 °C. It grows on rocky bottoms where its holdfast can become established and can be found as large underwater forests, with plants rising to and growing along the surface, at times up to 20 m in length.

Laminaria species are harvested principally in Norway, Scotland, Ireland, and France. *L. hyperborea* grows on rocky seabeds, usually at depths from 2 to 15 m. In Norway, where this type of seaweed is particularly abundant, fresh seaweed is harvested with specially designed equipment before being used for alginate production. *L. digitata* is found on either side of the low water mark and is usually harvested by hand when the plants are exposed at low tide. It is collected in France, Norway, and Scotland but the quantities are small in comparison with *L. hyperborea*. In France, it is harvested using small boats with a hydraulic arm fitted with a hook device at the end, which is lowered into the bed of *L. digitata* and rotated so that the weed wraps around it. The arm is then raised to the surface, bringing the seaweed with it.

Ascophyllum nodosum grows in the intertidal zone. It has been harvested by hand in Scotland and Ireland for more than a century. A mechanized harvesting technique was developed in Norway, whereby the seaweeds are cut and then pumped through a large diameter pipe into a net bag on a shallow-draught water jet-propelled vessel. The operation is carried out at high tide, and the bags can be left floating for later collection.

Other types of brown seaweeds for alginate production include *Lessonia*, *Ecklonia*, and *Durvillaea* species. *Lessonia* is collected in Chile, where it is cast up after storms. This particular species of brown seaweed has been popular in the alginate industry for the production of food-grade alginate due to its relatively high content of guluronic acid. *Lessonia trabeculata* grows in the sublittoral at a depth of 1–20 m. It has a very thick holdfast and stands up underwater, rather like *L. hyperborea*. Figure 1.4 shows an illustration of the brown seaweed *L. trabeculata*.

Figure 1.4 An illustration of the brown seaweed *Lessonia trabeculata*.

Ecklonia cava grows in deep water up to 20 m and is harvested by divers in both Japan and Korea. *Eisenia bicyclis* grows in a similar location and is collected along with *Ecklonia* in Japan. *Ecklonia* that has been cast up by storms is collected in Korea and South Africa. In Korea, it is used by the local alginate producer. The Korean industry also uses waste *Undaria* that is unsuitable for food uses, just as the Japanese industry uses similar waste from *S. japonica* seaweeds.

The alginate obtained from *Sargassum* and *Turbinaria* has a poor viscosity, so these species are used only when the above colder water species are not available. The Indian alginate industry is based on *Sargassum* that grows in the south, for example, along the coasts of Kerala and Tamil Nadu states. The species that grow in the north give a low-viscosity alginate, unsuitable for the main Indian market of textile printing. *Turbinaria* is used only when supplies of *Sargassum* are unavailable. The Philippines has large resources of *Sargassum*, but this is exported mainly to Japan for use in animal feeds and fertilizers.

Sargassum species are found worldwide in both the eulittoral and upper sublittoral zones. They exhibit a wide variety of shapes and forms. The alginate content is usually low, and the quality of the alginate is poor. For alginate extraction, they are regarded as the raw material of last resort.

None of the common seaweeds for alginate production are cultivated. They cannot be grown by vegetative means but must go through a reproductive cycle involving an alternation of generations. For alginate production, this makes cultivated brown seaweeds too expensive when compared to the costs of harvesting and transporting wild seaweeds. The only exception is *S. japonica* (formerly known as *Laminaria japonica*), which is now widely cultivated in China for food but is also used for alginate production. Since its initial development in the 1950s, the cultivation of *S. japonica* has been very successful in China, reaching about eight million tons of wet seaweed annually, of which about two-thirds is used as food and the rest is available for alginate production. Figure 1.5 shows an illustration of the cultivation of brown seaweed, *S. japonica*.

Figure 1.5 An illustration of the cultivation of brown seaweed, *Saccharina japonica*.

1.3 The Extraction of Alginate from Brown Seaweeds

1.3.1 General Description of the Extraction Process

Figure 1.6 shows an illustration of the wet and dry structures of brown seaweed, where alginic acid is the main structural component, accounting for up to 40% of the dry seaweed biomass. Alginate exists mainly in the intercellular mucilage and algal cell wall as a water-insoluble mixture of calcium, magnesium, potassium, and sodium salts, which provide the mechanical strength and flexibility of the seaweed as a marine bio-organism. In addition, as a hydrophilic biopolymer, alginate

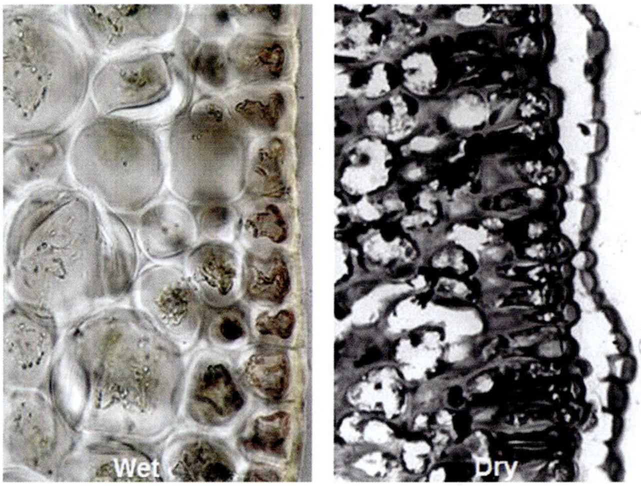

Figure 1.6 An illustration of the wet and dry structure of brown seaweed.

acts as a water reservoir, preventing dehydration when part of the seaweed is exposed to air. In general, the biological role and morphophysiological properties of alginate in brown seaweeds are similar to those of cellulose and pectin in terrestrial plants.

For the commercial extraction of alginate from brown seaweeds, F C Thornley first established a business based on using alginate as a binder for anthracite dust in 1923, and when that was not successful, he moved to San Diego. By 1927, his company was producing alginate for use in sealing cans. After some difficulties, the company changed its name to Kelp Products Corp., and in 1929, it was reorganized as Kelco Company.

Prior to the establishment of Kelco, there were a few companies established in the United Kingdom following the discovery of alginate by Stanford in 1881, such as British Algin Company Ltd. (1885), Blandola Ltd. (1908), Liverpool Borax Ltd. (1909). In 1934, Cefoil Ltd. was established to extract alginate from seaweeds in order to make fibers for military uses [18, 31]. World War II stimulated the alginate industry when production units were set up in Scotland and California using local seaweed resources of wrack and kelp. After the war, other production units followed suit and were constructed close to natural seaweed beds in the United States, Norway, France, the United Kingdom, Japan, and, more recently, China. The raw materials are mainly *M. pyrifera* in California, *L. hyperborea* in Norway, *L. digitata* in France, and *A. nodosum* in Scotland. In China, the kelp *S. japonica* was introduced from Japan and has been successfully cultivated on a large scale, usually grown on ropes along the Pacific coast.

Yields of alginate from different types of brown seaweeds vary greatly [24]. It has been reported that the yields of alginate as a percentage of dry seaweed biomass are, respectively, 18–45% for *M. pyrifera* [8, 30], 16–36% for *L. digitata* [4, 16], 14–21% for *L. hyperborea* [4], 17–25% for *S. japonica* [12], 16–34% for *Saccharina latissima* [16], 13–29% for *L. trabeculata* [5], 24–28% for *Ecklonia arborea* [9], 45–55% for *Durvillaea potatorum* [15, 23] and 12–16% for *A. nodosum* [22].

Figure 1.7 shows a process flow chart for the extraction of alginate from brown seaweeds. During the extraction process, the goal is to obtain dry, powdered sodium alginate. The natural calcium and magnesium salts of alginic acid in the seaweed biomass do not dissolve in water, while the sodium salt does. Therefore, the rationale behind the extraction of alginate from the seaweed is to convert all the alginate salts to sodium salts and dissolve them in water. After removing the seaweed residue by filtration, alginic acid can be recovered from the aqueous solution.

During the extraction process, once the alginate component of seaweed is in the aqueous extraction medium, there are two different ways to recover it. The first is to add acid to the extraction solution to convert sodium alginate into alginic acid, which does not dissolve in water and hence can be separated from the water. The alginic acid separates as a soft gel, and some of the water must be removed from this. After this has been done, alcohol is added to the alginic acid, followed by sodium carbonate, which converts the alginic acid into sodium alginate. Since sodium alginate does not dissolve in the mixture of alcohol and water, it can be separated from the mixture, dried, and milled to an appropriate particle size.

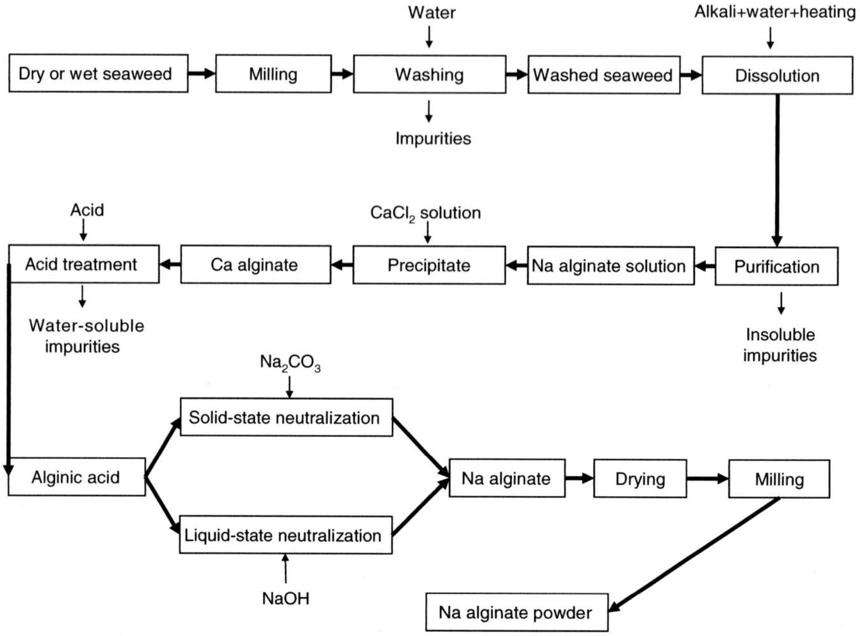

Figure 1.7 Process flow chart for the extraction of alginate from brown seaweeds. Source: Adapted with permission from fig. 3.4, Qin [26].

The second method of recovering sodium alginate from the initial extraction solution is to add a calcium salt, which results in the formation of water-insoluble calcium alginate that can be separated from the aqueous medium. In order to further purify the alginate material, acid is added to convert it into alginic acid before adding sodium carbonate to convert alginic acid to sodium alginate, which is extruded into pellets that are then dried and milled.

These two processes are straightforward, and the chemistry is simple, i.e. convert the water-insoluble alginate salts in the seaweed into soluble sodium alginate and precipitate it either in the form of alginic acid or calcium alginate so that alginate can be separated from the extract solution. The difficulties lie in handling the materials encountered in the process. In order to extract alginate, the seaweed biomass should be broken into pieces and stirred with a solution of an alkali, usually sodium carbonate. Over a period of time, alginate dissolves as sodium alginate to produce a thick slurry, which also contains parts of the seaweed that do not dissolve, mainly cellulose. The solution is often too viscous to be filtered and must be diluted with a very large quantity of water before it passes through a filter cloth. However, the pieces of undissolved residue are very fine and can quickly clog the filter cloth. Therefore, before filtration is started, a filter aid, such as diatomaceous earth, must be added, which holds most of the fine particles away from the surface of the filter cloth and facilitates filtration. Since the filter aid is expensive and in order to reduce the quantity of filter aid needed, some processors force air into the extract as it is being diluted with water, where fine air bubbles attach themselves to the particles of residue. During this time, the diluted extract is left standing for several hours while the air rises to the top, taking the residue particles with it. This frothy mix of air and residue is then removed from the top, and the solution is withdrawn from the bottom and pumped to the filter. The next step is the precipitation of the alginate from the filtered solution, either as alginic acid or calcium alginate.

1.3.2 A Comparison of Alginic Acid Method and Calcium Alginate Method

The alginic acid method uses acid to convert alginate in the filtered extract into alginic acid, which is water insoluble in the form of soft and gelatinous pieces. This jelly-like mass of alginic acid contains only 1–2% alginic acid and 98–99% water. At this stage, it is too soft to allow the use of a screw press. By centrifuging, the solid content can be increased to 7–8% before alcohol (usually ethanol or isopropanol) is added to produce a 50 : 50 mixture of alcohol and water. Solid sodium carbonate is then added gradually until the resulting paste reaches the desired pH. The paste of sodium alginate can be extruded as pellets or oven dried and milled.

In the calcium alginate method, calcium alginate is formed when calcium chloride is added to the filtered extract. The resultant calcium alginate precipitate can be readily separated and washed with water to remove excess calcium. It is then stirred in dilute acid and converted to alginic acid, which can be squeezed in a screw press to remove excess water. The product coming out of the screw press contains about 20–25% alginic acid, which is then mixed with sodium carbonate to convert it into sodium alginate.

The advantage of the calcium alginate method is that calcium alginate can be precipitated in a fibrous form that can be readily separated and then converted into alginic acid, which is still fibrous and can be readily separated. In addition, the final sodium alginate product may contain some calcium ions, which help control the viscosity of the final product. The alginic acid method saves one step, i.e. the formation of calcium alginate; however, when alginic acid is precipitated in this process, it forms a gelatinous precipitate, which is very difficult to separate, and the overall losses of alginic acid are generally greater than in the calcium alginate method. It is also more difficult to remove water from within the gel structure of the separated alginic acid. The water content in the dewatered alginic acid is often high so that alcohol must be used as a solvent for the conversion to sodium alginate, which makes the process more expensive.

1.3.3 Process Control

During the production of alginate from brown seaweeds, appropriate process control is needed for color control of the product, water supply, and waste disposal. If the original seaweed is highly colored, such as with *Ascophyllum* seaweeds, the alkaline extract will also be highly colored, and a dark product is produced, which is not suitable for high-value applications such as medical devices. Lighter colored seaweeds, such as *Macrocystis*, yield light-colored alginate suitable for food and other applications.

Sodium hypochlorite can be used to bleach alginate in the filtered alkaline extract or during the final conversion stage. As an oxidant, the excessive use of sodium hypochlorite can lower the molecular weight of alginate and reduce its value. An alternative method to reduce color is to apply formalin in the extraction solution, whereby colored compounds are bound to cellulose in the seaweed cell walls so that much of the color is left behind in the seaweed residue when the alkaline extract is filtered.

The extraction of alginate from brown seaweeds requires a large amount of processing water, in particular when the viscous alkaline extract is diluted to a low viscosity suitable for filtration. Up to 1000 tons of water are required to produce 1 ton of alginate, and a reliable source of water supply is important for alginate production. In addition, a large amount of wastewater also needs to be treated and preferably recycled. It should be pointed out that the wastewater from the alginate extraction process is relatively harmless, and in some countries, the waste is pumped out to sea. Where environmental concerns are greater, or when water supplies are limited, recycling is not too difficult and its costs may be partly offset by the lowering of the quantity and cost of water used by the factory.

In addition to wastewater, there are also solid wastes in the form of seaweed residues, which contain cellulose, protein, minerals, and other organic and inorganic components. The residues are rich in bioactive substances, and they are now commonly used as raw materials for the production of fertilizers.

1.3.4 Key Process Parameters

The extraction of alginate from brown seaweeds involves the dissolution of alginic acid in an alkaline solution and its precipitation with an acid or calcium salt. Although the chemistry is simple, a number of process parameters need to be addressed and controlled within appropriate ranges. Some of the key process parameters are discussed below.

1.3.4.1 Size Reduction of Raw Materials

The seaweeds that are fed into the extraction process can be in many forms. They can be freshly collected from the sea such as those used in Norway, while those collected in Chile are usually dried and broken up into small pieces of about 5–10 mm square, which makes it easy to wet the seaweed when they are soaked in water at the beginning of the extraction process. It also helps to facilitate the penetration of formalin, acid, and alkali more thoroughly and more rapidly. In addition, process flow is also smoother when the seaweeds are in small pieces that can be pumped and transported through the process pipeline. Figures 1.8 and 1.9 show a whole plant of dried seaweed and broken pieces of dried brown seaweed, respectively.

1.3.4.2 Acid Treatment

Alginate exists in mixed calcium, magnesium, potassium, and sodium salts in the seaweed biomass, and it has been shown that a pretreatment with dilute mineral acid can remove the metal ions, leading to a more efficient extraction process. In the

Figure 1.8 The author holds a whole plant of dried seaweed.

(a) (b)

Figure 1.9 An illustration of the broken pieces of dried brown seaweed. (a) Seaweeds being sun dried on the beach and (b) broken pieces of brown seaweed.

pretreatment, the alginate is first converted into alginic acid before being converted into sodium alginate following the addition of alkali (usually sodium carbonate), as illustrated below:

$$\mathrm{Ca(Alg)_2 + 2H^+ = 2HAlg + Ca^{++}}$$

$$\mathrm{HAlg + Na^+ = NaAlg + H^+}$$

During the acid treatment, all the acid-soluble phenolic compounds are removed, which is important in that the phenolic compounds can form brown oxidation/polymerization products with alkali and are largely responsible for the brown discoloration, which occurs during alkaline extraction. Hence, pretreatment of the seaweed with acid before alkaline extraction gives a more efficient extraction, a less colored product, and a reduced loss of viscosity during extraction. During the treatment, the seaweed is stirred with 0.1 M sulfuric acid or hydrochloric acid for 30 minutes, with treatment temperature ranging from room temperature to about 50 °C depending on the seaweed used. Little degradation of alginate occurs with most species of seaweed at temperatures up to 40–50 °C. At the end of the treatment, the slurry of seaweed and acid solution can be separated on a rotary drum screen.

1.3.4.3 Formaldehyde Treatment

The discoloration of alginate can be further reduced by pretreatment with formaldehyde. It has been found that phenolic compounds and formaldehyde can react to form insoluble products, which remain in the seaweed residue during the extraction process. In practice, a 0.1–0.4% commercial formalin solution is used to treat the seaweed, usually at room temperature for 15–30 minutes. After the treatment, the seaweed is separated using a rotary drum screen, and the solids are used in the alkaline extraction.

1.3.4.4 Alkaline Extraction

The addition of alkali to the extraction solution converts alginate to a soluble form that can be separated from the rest of the seaweed biomass. Sodium carbonate (soda ash) is usually used as the alkali because of its low cost. During the production process, the seaweed is stirred in a tank with the sodium carbonate solution at about 1.5% concentration. When the seaweed has undergone size reduction and acid pretreatment, a good extraction can be achieved for two hours at 50 °C with little degradation of alginate, although the time can be reduced by using higher temperatures, usually with some loss of viscosity in the final product. As extraction proceeds, the extract becomes thicker, requiring dilution during the subsequent processes.

1.3.4.5 Separation of Alginate from Insoluble Seaweed Residue

The separation of alginate from insoluble seaweed residue involves several steps, including flotation, filtration, precipitation, bleaching, conversion, dewatering, drying, and milling.

Flotation When sodium alginate is dissolved in the extract solution, a viscous mixture is formed where the alkali-insoluble seaweed residue is usually slimy and finely divided and can rapidly clog filter cloths if conventional filtration methods are used. Although some of the residues can be removed using centrifuges, the clarity of the resulting solution is usually poor. During commercial production, the major portion of the insoluble residue is usually removed by a flotation process, where the extract is first diluted with 4–6 times its volume of water to produce a suitable viscosity of about 25–100 cps. A small quantity of flocculant is then added, and air is forced into the liquid. After being left to stand for several hours, the fine particles of insoluble residue are raised to the surface by the rising air bubbles and are scraped from the surface and the clarified liquor beneath it. Since the cellulose residue has a negative charge, cationic flocculants such as polyacrylamides and chitosan are often used.

Filtration After the flotation process, the remaining insoluble residue is filtered to give the final product good quality. Because the residue is very fine, filter cloths are rapidly blocked and the best method is to use a rotary precoat vacuum filter where the rotating drum of the filter is coated with a 2–3 cm layer of precoat material, preferably perlite because it gives a more porous medium than diatomaceous earth and so does not block as easily. During filtration, a blade on the rotary filter continually removes the top surface of the precoat so that a clean filter surface is always available. After 9–10 hours, most of the precoat has been removed by the scraper, filtration is stopped, and a new layer of precoat is deposited. Great care is necessary in selecting the appropriate grade of perlite and the correct cloth to support the precoat medium.

Precipitation of Calcium Alginate The purified alginate after filtration is a dilute aqueous solution of sodium alginate, and because of the presence of a large percentage of water, it is not economical to dry it with conventional methods such as heat evaporation. It must be precipitated out either as calcium alginate or as alginic acid. The process is normally carried out by adding the diluted extract to a 10% calcium chloride solution, resulting in fibrous precipitate that can be easily handled on a metal screen. If calcium chloride solution is added to the extract, a soft gel is formed, which is more difficult to process further. After washing with water, the fibrous calcium alginate is then treated with dilute mineral acid when the Ca^{2+} ions are exchanged for H^+ ions to produce fibrous alginic acid, which is dewatered using a screw press. The dewatered product should contain at least 25% solids if it is to be used in the conversion process.

Bleaching For food- and medical-grade alginate, it is necessary to use bleaching for the improvement of color and odor. In order to avoid degradation, the bleaching process is best carried out with calcium alginate since it is chemically more stable than alginic acid. The bleaching process is normally carried out by adding a 12% sodium hypochlorite solution to a suspension of calcium alginate in water.

Conversion of Alginic Acid to Sodium Alginate Alginic acid is converted into sodium alginate in two ways. In the dry method, the dewatered alginic acid is mixed with solid alkali, usually sodium carbonate, in a mixer suitable for blending heavy pastes. A thick paste is formed when the resultant sodium alginate is dissolved in the limited amount of water in the system. If the alginic acid content is less than 25%, the resulting paste may be too fluid for the next step when the paste is forced through small holes and the extrusions are chopped into pellets and dried. The dried pellets with about 10% moisture can be milled to an appropriate particle size, usually about 60 mesh or 250 μm. In the wet conversion method, NaOH is dissolved in ethanol before reacting with alginic acid to produce sodium alginate.

1.4 Ultrapure Alginate

Alginate obtained in the above-described procedure contains several mitogens and cytotoxic impurities, making it unsuitable for biomedical applications. Ultrapure and amitogenic alginates suitable for biomedical purposes are prepared by using more rigorous extraction processes. For example, free flow electrophoresis was applied as one technique to remove mitogenic impurities from commercial alginates [32]. This method was, however, not suitable for large-scale processing because it was time consuming and required expensive electrophoresis equipment. A chemical extraction method was therefore described by using Ba-alginate gels [14]. Ba^{2+} ions show higher affinity toward alginates compared to Ca^{2+} ions. Ba-alginate gels are stable in acidic and neutral pH environments but disintegrate under alkaline pH conditions. During the purification process, mitogenic contaminants are first eluted from Ba-alginate beads by treatment with various solutions followed by ethanol extraction, after which the pure alginate beads are dissolved in an alkaline

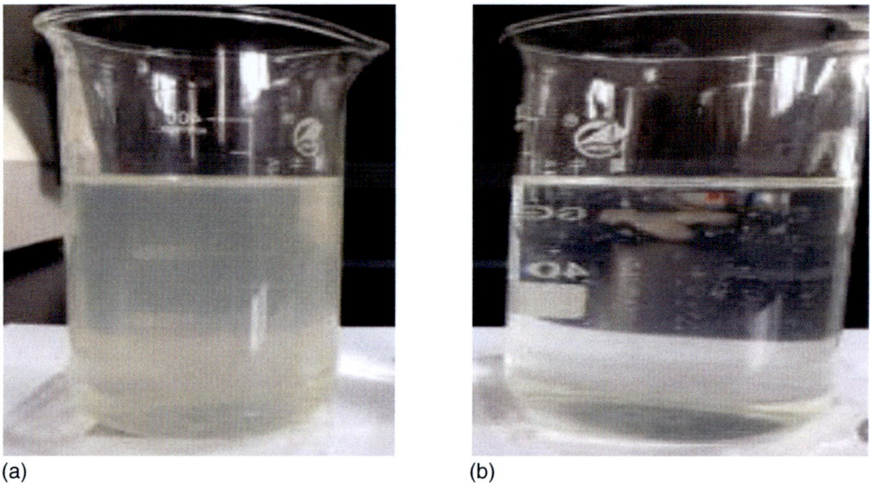

(a) (b)

Figure 1.10 Alginate solutions (a) before and (b) after purification.

solution, which is then dialyzed. Once Ba^{2+} ions are fully exchanged for Na^{+} ions, the purified sodium alginate is precipitated from the solution by the addition of ethanol. Figure 1.10 shows alginate solutions before and after purification.

1.5 Summary

Many species of brown seaweed are commercially available from many parts of the world, which gives the alginate extract industry a wide variety of raw materials for the production of alginate with different chemical and physical properties. Some may yield an alginate that gives a strong gel, another a weaker gel; one may readily give a cream/white alginate, another may give that only with difficulty and is best used for technical applications where color does not matter. Alginate producers often prefer to buy a mixture of species of seaweed that allows them to blend their products to give them properties suitable for specific end uses.

References

1. Andresen, I.L., Skipnes, O., Smidsrod, O. et al. (1977). Some biological functions of matrix components in benthic algae in relation to their chemistry and the composition of seawater. *ACS Symp. Ser.* 48: 361–381.
2. Atsuki, K. and Tomoda, Y. (1926). Studies on seaweeds of Japan, I. The chemical constituents of Laminaria. *J. Soc. Chem. Ind. Jpn.* 29: 509–517.
3. Bird, G.M. and Haas, P. (1931). On the constituent nature of the cell wall constituents of *Laminaria* spp. mannuronic acid. *Biochem. J.* 25: 26–30.
4. Black, W.A.P. (1950). The seasonal variation in weight and chemical composition of the common British Laminariaceae. *J. Mar. Biol. Assoc. UK* 29: 45–72.
5. Chandia, N. (2001). Alginic acids in *Lessonia trabeculata*: characterization by formic acid hydrolysis and FT-IR spectroscopy. *Carbohydr. Polym.* 46 (1): 81–87.
6. Fischer, F.G. and Dorfel, H. (1955). Die Polyuronsauren der Braunalgen (Kohlenhydrate der Algen). *Z. Phys. Chem.* 302: 186–203.
7. Gorin, P.A.J. and Spencer, J.F.T. (1966). Exocellular alginic acid from *Azotobacter vinelandii*. *Can. J. Chem.* 44: 993–998.
8. Hernandez-Carmona, G. (1985). *Variacion estacional del contenido de alginatos en tres especies de feofitas de Baja California Sur*, vol. 2, 29–45. Invest Marinas CICIMAR.
9. Hernnndez-Carmona, G. (1985). *Variación estacional del contenido de alginatos en tres especies de feofitas de Baja California Sur*, vol. 2, 29–45. Invest Marinas CICIMAR.
10. Hoefer, D., Schnepf, J.K., Hammer, T.R. et al. (2015). Biotechnologically produced microbial alginate dressings show enhanced gel forming capacity compared to commercial alginate dressings of marine origin. *J. Mater. Sci. Mater. Med.* 26: 162–171.

11 Hoek, C., Mann, D.G., and Jahns, H.M. (1995). *Algae: An Introduction to Phycology*, 165–218. Cambridge: Cambridge University Press.

12 Honya, M., Kinoshita, T., Ishikawa, M. et al. (1993). Monthly determination of alginate, M/G ratio, mannitol, and minerals in cultivated *Laminaria japonica*. *Nippon Suisan Gakkaishi* 59 (2): 295–299.

13 Indergaard, M. and Skjak-Braek, G. (1987). Characteristics of alginate from *Laminaria digitata* cultivated in a high phosphate environment. *Hydrobiologia* 151/152: 541–549.

14 Klöck, G., Frank, H., Houben, R. et al. (1994). Production of purified alginates suitable for use in immunoisolated transplantation. *Appl. Microbiol. Biotechnol.* 40: 638–643.

15 Lorbeer, A.J., Charoensiddhi, S., Lahnstein, J. et al. (2017). Sequential extraction and characterization of fucoidans and alginates from *Ecklonia radiata*, *Macrocystis pyrifera*, *Durvillaea potatorum*, and *Seirococcus axillaris*. *J. Appl. Phycol.* 29: 1515–1526.

16 Manns, D., Nielsen, M.M., Bruhn, A. et al. (2017). Compositional variations of brown seaweeds *Laminaria digitata* and *Saccharina latissima* in Danish waters. *J. Appl. Phycol.* 29: 1493–1506.

17 Miawa, T. (1930). Alginic acid. *J. Chem. Soc. Jpn.* 51: 738–745.

18 Moradali, M.F., Ghods, S., and Rehm, B.H.A. (2018). Alginate biosynthesis and biotechnological production. In: *Alginates and Their Biomedical Applications*, Springer Series in Biomaterials Science and Engineering, vol. 11 (ed. B.H.A. Rehm and M.F. Moradali), 1–25. Singapore: Springer Nature Singapore Pte Ltd.

19 Nelson, W.L. and Cretcher, L.H. (1929). The alginic acid from *Macrocystis pyrifera*. *J. Am. Chem. Soc.* 51: 1914–1918.

20 Nelson, W.L. and Cretcher, L.H. (1930). The isolation and identification of D-mannuronic acid lactone from the *Macrocystis pyrifera*. *J. Am. Chem. Soc.* 52: 2130–2134.

21 Nelson, W.L. and Cretcher, L.H. (1932). The properties of D-mannuronic acid lactone. *J. Am. Chem. Soc.* 54: 3406–3409.

22 Obluchinskaya, E.D., Voskoboinikov, G.M., and Galynkin, V.A. (2002). Contents of alginic acid and fuccidan in Fucus algae of the Barents Sea. *Appl. Biochem. Microbiol.* 38 (2): 186–188.

23 Panikkar, R. and Brasch, D.J. (1996). Composition and block structure of alginates from New Zealand brown seaweeds. *Carbohydr. Res.* 293 (1): 119–132.

24 Peteiro, C. (2018). Alginate production from marine macroalgae, with emphasis on kelp farming. In: *Alginates and Their Biomedical Applications*, Springer Series in Biomaterials Science and Engineering, vol. 11 (ed. B.H.A. Rehm and M.F. Moradali). Singapore: Springer Nature Singapore Pte Ltd.

25 Porse, H. and Rudolph, B. (2017). The seaweed hydrocolloids industry: 2016 updates, requirements and outlook. *J. Appl. Phycol.* 29: 1–24.

26 Qin, Y. (2018). *Bioactive Seaweeds for Food Applications*. San Diego: Academic Press.

27 Schmidt, E. and Vocke, F. (1926). Zur kenntnis der Poly-glykuronsauren. *Chem. Ber.* 59: 1585–1588.

28 Stanford, E.C.C. (1881). *Improvements in the manufacture of useful products from seaweeds*. British Patent 142.

29 Stanford, E.C.C. (1883). New substance obtained from some of the commoner species of marine algae: algin. *Chem. News* 47: 254–257.
30 Westermeier, R., Murua, P., Patino, D.J. et al. (2012). Variations of chemical composition and energy content in natural and genetically defined cultivars of *Macrocystis* from Chile. *J. Appl. Phycol.* 24: 1191–1201.
31 Woodward, F. (1951). The Scottish seaweed research association. *J. Mar. Biol. Assoc. UK* 29 (3): 719–725.
32 Zimmermann, U., Klöck, G., Federlin, K. et al. (1992). Production of mitogen-contamination free alginates with variable ratios of mannuronic acid to guluronic acid by free flow electrophoresis. *Electrophoresis* 13: 269–274.

Further Reading

Black, W.A.P. and Woodward, F.N. (1954). Alginates from common British brown marine algae. In: *Natural Plant Hydrocolloids*, Advances in Chemistry, vol. 11, 83–91. American Chemical Society.
Clark, D.E. and Green, H.C. (1936). Alginic acid and process of making same. US Patent 2,036,922.
Glicksman, M. (1969). *Gum Technology in the Food Industry*, 67–68. New York: Academic Press.
Grasdalen, H., Larsen, B., and Smidsrod, O. (1970). A.P.M.R. study of the composition and sequence of uronate residues in alginates. *Carbohydr. Res.* 68: 23–31.
Grasdalen, H., Larsen, B., and Smidsrod, O. (1981). ^{13}C-NMR studies of monomeric composition and sequence in alginate. *Carbohydr. Res.* 89: 179–191.
Haug, A. (1967). Correlation between chemical structure and physical properties. *Acta Chem. Scand.* 21: 768–778.
Haug, A. and Larsen, B. (1962). Quantitative determination of the uronic acid composition of alginates. *Acta Chem. Scand.* 16: 1908–1918.
Haug, A., Larsen, B., and Smidsrod, O. (1966). A study of the constitution of alginic acid by partial acid hydrolysis. *Acta Chem. Scand.* 20: 183–190.
Haug, A., Larsen, B., and Smidsrod, O. (1974). Uronic acid sequence in alginate from different sources. *Carbohydr. Res.* 32: 217–225.
Hilton, K.A. (1972). Sodium alginate in the textile industry. *Colourage* 19 (10): 65–68.
Ji, M.H. (1984). Studies on the M : G ratios in alginate. *Hydrobiologia* 116/117: 554–556.
McHugh, D.J. (2003). A guide to the Seaweed Industry, FAO Fisheries Technical Paper 441. Rome: FAO
Morris, E.R., Rees, D.A., and Thom, D. (1980). Characteristics of alginate composition and block-structure by circular dichroism. *Carbohydr. Res.* 81: 305–314.
Penman, A. and Sanderson, G.R. (1972). A method for the determination of uronic acid sequence in alginates. *Carbohydr. Res.* 25: 280–286.
Stanford, E.C.C. (1881). Improvements in the Manufacture of Useful Products From Seaweeds. British Patent 142.
Thornley, F.C. and Walsh, M.J. (1931). Process of preparing alginic acid and compounds thereof. US Patent 1,814,981.

2

Chemical, Physical, and Biological Properties of Alginic Materials

2.1 Introduction

Alginate is a linear copolymer composed of blocks of (1–4)-linked β-D-mannuronate (M) and its C-5 epimer α-L-guluronate (G) residues, covalently linked together in different sequences or blocks, where the monomers can appear in homopolymeric blocks of consecutive M-residues (M-blocks), consecutive G-residues (G-blocks), or alternating M- and G-residues (MG-blocks). Alginate is generally regarded as a water-soluble polymer, commonly used in the form of sodium alginate. However, in its various fields of applications, alginate can be made into a variety of end products with different chemical and physical structures, with the resultant materials possessing a large number of unique properties.

In order to fully apprehend the properties and applications of alginate as a polymeric material, this chapter gives an analysis of the chemical, physical, and biological properties of alginate.

2.2 The Chemical Structure of Alginic Acid

2.2.1 Early Studies and Basic Structural Feature

Although the discovery of alginate dates back to 1881, it took several decades to completely elucidate its chemical structure. It was not until 1926 when the presence of uronic acids, specifically mannuronic acid (M), was found as a structural feature of the alginate backbone [1, 41] and the glycosidic bonds connecting C-1 of one mannuronic unit to C-4 of the following one were found to display a ß-configuration, similar to the structure of cellulose. Fisher and Dörfel [9] found that another uronic acid, i.e. α-L-guluronic acid (G), was also present in alginate. The discovery of this chemical inhomogeneity helps to explain the variations of physical properties of alginate extracted from different seaweeds, since the content of the two different uronic groups varies over a wide range depending on the algal source and the part of the plant.

Haug et al. [20] carried out partial hydrolysis of alginates followed by fractionation to obtain alginates containing different copolymer compositions, either

Alginate Fibers and Wound Dressings: Seaweed Derived Natural Therapy, First Edition. Yimin Qin.

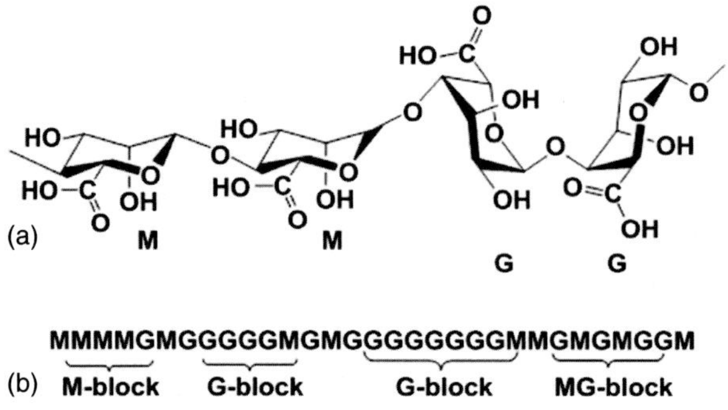

Figure 2.1 Polymeric structure of alginate, (a) Chemical structures of mannuronic acid and guluronic acid; (b) Sequence of M and G blocks along the polymeric chain. Source: Adapted with permission from Fig. 7.3, Qin [39].

mainly M-rich or mainly G-rich residues. It was then proposed that alginate is a copolymer consisting of M-blocks, G-blocks, and M/G alternating blocks [22]. This copolymer structure was further illustrated by computer-driven mathematical models [28, 36, 48].

Following the early studies on the chemical nature of alginic acid, the development of advanced instrumental analysis techniques made it possible to offer a much clear understanding of the composition and backbone structure of alginic acid as a copolymer. Penman and Sanderson [37] used ^{1}H- and ^{13}C-NMR spectroscopy to determine the M/G ratio in alginates from the peak ratios obtained in the ^{1}H-NMR spectrum. A later study by Grasdalen et al. [13] found that ^{1}H NMR can quantitatively determine the fractions of the four dyads, i.e. MM, MG, GM, and GG. As shown in Figure 2.1, it is now commonly accepted that alginate is a linear binary copolymer of 1–4-linked M and G residues arranged in a blockwise pattern with homopolymeric regions of G residues (G-blocks) and homopolymeric regions of M residues (M-blocks) interspersed by regions where the two monomers coexist in a strictly alternating sequence (MG-blocks).

2.2.2 M/G Ratio and Distribution

Mannuronic acid and guluronic acid are C-5 epimers. Stereochemically, because an M block is formed from equatorial groups at C-1 and C-4, it is relatively straight, like a flat ribbon, while the G block is formed from axial groups at both C-1 and C-4, resulting in a buckled chain structure that allows it to have the physical space to contain metal ions essential for gel formation, for example, the combination of calcium ions with alginic acid involves the G blocks, with a high proportion of G-block corresponding to a great gel strength.

The industrial utilization of alginate places specific requirements on the G/M composition, and it is important to quantify the G/M ratio of alginate as a raw material. Conventional analytical methods of chemical hydrolysis [21] and more advanced methods such as NMR [14] have been used to determine the MM, GG, and MG contents of alginate extracted from different species of brown seaweeds, which show considerable variations, resulting in the differences of physical properties of gels, films, fibers, and other products made from alginate. Table 2.1 summarizes the G, M, GG, MM, and MG/GM contents of alginate from different sources.

Table 2.1 G, M, GG, MM, and MG/GM contents of alginate from different sources.

Type of brown seaweeds	F_G (%)	F_M (%)	F_{GG} (%)	F_{MM} (%)	$F_{MG,GM}$ (%)
Saccharina japonica	35	65	18	48	17
Laminaria digitata	41	59	25	43	16
Laminaria hyperborea, blade	55	45	38	28	17
Laminaria hyperborea, stipe	68	32	56	20	12
Laminaria hyperborea, outer cortex	75	25	66	16	9
Macrocystis pyrifera	39	61	16	38	23
Ascophyllum nodosum, fruiting body	10	90	4	84	6
Ascophyllum nodosum, old tissue	36	64	16	44	20
Lessonia nigrescens	38	62	19	43	19
Ecklonia maxima	45	55	22	32	32
Durvillaea antarctica	29	71	15	57	14

There have been many studies on the effect of G, M, GG, MM, and MG/GM contents on the physical properties of alginates [13, 22, 33, 46], which demonstrated the impact of chemical composition on physical performances. However, the G/M composition is mainly controlled by nature in the sense that different types of seaweeds and different parts of a particular seaweed plant contain alginate with different G and M contents. It is in part the result of evolution, whereby seaweeds can regulate their physical properties in the sense that a high G content corresponds to a high stiffness of the plant body, while a high M content is associated with a soft seaweed structure, which is important to survive in a rough sea. The G/M ratio is related to the types of seaweeds, the growth stage, the weather conditions where the seaweeds grow, etc. Since alginates extracted from different parts of the same seaweed plant have different G/M contents, for example, the G contents for alginate extracted from the blade, stipe, and outer cortex are 55%, 68%, and 75%, respectively, it is feasible to cut seaweeds into separate parts to produce alginates with either high G content or high M content.

2.2.3 C-5 Epimerization and Designer Alginate

When synthesized naturally both in seaweed and bacteria, alginates are first produced through the polymerization of mannuronan, followed by the enzymatic inversion of the configuration at the C-5 position of part of the uronic acids into guluronic acids, without breaking the glycosidic bond. It was found that seven epimerases were encoded in the genome of the bacterium *Azotobacter vinelandii*, which are named "AlgE1" to "AlgE7." These enzymes have been sequenced, cloned, and overexpressed in *Escherichia coli* strains [7, 8, 58]. It was then possible to obtain these seven epimerases from the bacteria cells by sonication, centrifugation, filtration, and ion-exchange chromatography. When used on alginate extracted from seaweeds, it is possible to alter the G/M ratio through enzymatic treatment with these epimerases.

It is interesting to note that each of the seven recombinant mannuronan C-5 epimerases generates specific nonrandom epimerization patterns, thus allowing the introduction of G residues into the alginate chains in different block structures. For example, the C-5 epimerase AlgE4 is able to convert homopolymeric M-blocks into strictly alternating copolymeric sequences, while the AlgE6 epimerase is a G-block-forming enzyme [15, 16, 23, 24]. Through the selective use of the different epimerases, alginate with tailored G/M sequences can be engineered, which cannot be obtained from natural sources. For example, the use of epimerase technology has made it possible to generate alginate samples composed of almost exclusively one block sequence [4, 31]. By converting β-(1→4)-linked D-mannuronic acid residues to α-(1→4)-linked L-guluronic acid residues in the polymer chain, the resulting alginate forms stronger gels that are required in high-valued applications, such as food and biomedical materials [44].

2.2.4 Molecular Weight and Distribution

The molecular weight of alginate extracted from either seaweeds or bacterial sources is polydisperse and varies widely. This is due to two main reasons. First, the production of polysaccharides is not gene-encoded but is under enzymatic control. Second, the extraction process causes substantial depolymerization of natural alginate. Commercial alginates have a weight–average molecular weight of approximately 200 000, although naturally occurring alginates with molecular weights as high as 400 000–500 000 are available. As a polymeric material, alginate typically has a polydispersity index ranging from 1.5 to 3. Multiangle laser light scattering coupled with size-exclusion chromatography using a refractive index detector was found to be an efficient and reliable method for the determination of both the molecular weight and its distribution [60].

2.2.5 Chemical Stability

As a polysaccharide, alginate undergoes hydrolytic cleavage under acidic conditions. The mechanism of acid hydrolysis of the glycosidic bond has been described by Timell [59], which involves three steps: (1) protonation of the glycosidic oxygen to give the conjugate acid; (2) heterolysis of the conjugate acid, forming a nonreducing end group and a carbonium–oxonium ion; and (3) rapid addition of water to the carbonium–oxonium ion, forming a reducing end group. Figure 2.2 illustrates the mechanism of acid hydrolysis of alginate.

Alginic acid degrades more rapidly than sodium alginate or calcium alginate. The reason for this enhanced degradation rate is thought to be intramolecular catalysis by the C-5 carboxyl groups [52]. On the other hand, under neutral conditions, sodium alginate in the form of dry powder can be stored without degradation in a cool and dry place and away from sunlight for several months. The shelf life can be extended to several years by storing it in the freezer. Tables 2.2 and 2.3, respectively, show the

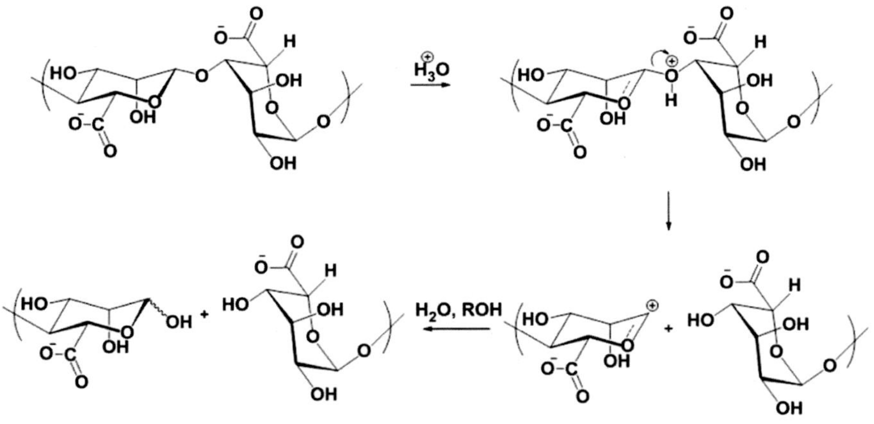

Figure 2.2 Mechanism of acid hydrolysis of alginate.

Table 2.2 Viscosity changes of sodium alginate when stored for a year.

Viscosity grade	Original viscosity of 1% solution (mPa s)	Storage temperature (°C)	Viscosity of 1% solution after one year (mPa s)
Low	42	0	40
		25	39
		35	34
Medium	470	0	450
		25	410
		35	240
High	1300	0	1200
		25	580
		35	260

Original data: Technical center of Qingdao Brightmoon Seaweed Group.

viscosity changes of sodium alginate and other alginate products when stored for a year.

In addition to acids or bases, alginates are also susceptible to chain degradation at neutral pH when in the presence of reducing compounds. A number of reducing compounds such as hydroquinone, sodium sulfite, sodium hydrogen sulfide, cysteine, ascorbic acid, hydrazine sulfate, and leuco-methylene can cause degradation, in addition to phenolic compounds which are naturally present in brown seaweeds and can be combined with alginate. The mechanism of degradation involves the formation of a peroxide leading to free radical creation, which eventually causes breakdown of the alginate chain [50, 51, 53]. Sterilization techniques such as heat treatment, autoclaving, ethylene oxide treatment, and γ-irradiation can also cause alginate degradation [29].

It is sometimes necessary to breakdown the polymeric chain of alginate in order to produce low molecular grade alginate or oligomers. In these cases, it may be

Table 2.3 Viscosity changes of alginate products when stored for a year.

Products	Viscosity of 1% aqueous solution (mPa s)	
	At the beginning	After storing for a year
Ammonium alginate	1400	650
Potassium alginate	300	275
Propylene glycol alginate	150	107
	420	253
Sodium alginate	37	35
	260	210
	580	460
	1200	590

Original data: Technical center of Qingdao Brightmoon Seaweed Group.

useful to treat alginate with oxidants such as sodium periodate [27, 47, 60]. In addition, in vivo biodegradation of alginates is performed by endolyases, which catalyze the splitting of the 1–4 glycosidic bond via a ß-elimination reaction. Endolyases are widely distributed in nature in prokaryotic and eukaryotic microorganisms and in bacteriophages that use alginate as a source of carbon. Lyases are found in the bacterial species producing alginates, such as *A. vinelandii* and *Pseudomonas aeruginosa*, but not in the human gastrointestinal tract. Alginate-degrading enzymes have also been isolated from marine algae, including *Laminaria digitata*. Like epimerases, endolyases exhibit sequence specificity in cleavage of either M or G residues in alginate chains [10, 61].

2.3 Physical Properties of Alginic Materials

The degree of polymerization (DP) is a measure of the average molecular weight of the alginate molecules directly related to the viscosity of alginate solutions. As shown in Table 2.4, sodium alginate is commercially produced in three main grades, usually described as low, medium, and high viscosity alginates, measured by the viscosity of its 1% aqueous solution.

Alginates with a high DP are less stable than those with a low DP. Low-viscosity sodium alginates (up to about 50 mPa·s) have been stored at 10–20 °C with no

Table 2.4 Molecular weight and degree of polymerization of alginate.

Grades	Molecular weight	Degree of polymerization
Low viscosity	12 000–80 000	60–400
Medium viscosity	80 000–120 000	400–600
High viscosity	120 000–190 000	600–1000

Original data: Technical center of Qingdao Brightmoon Seaweed Group.

observable change in molecular weight over three years. Medium-viscosity sodium alginates (up to about 400 mPa·s) show a 10% loss at 25 °C and a 45% loss at 33 °C after one year. Higher viscosity alginates are less stable. It is, therefore, important to store them in a cool place of 25 °C or lower, since elevated temperatures can cause significant depolymerization which affects commercially useful properties such as gel strength.

The monovalent cation salts of alginic acid with Na^+, K^+, NH_4^+, and $(CH_2OH)_3NH^+$ are soluble in water. Three methods can be used to dissolve alginates. In high-shear mixing, the principle is to prevent the clumping together of the particles, which become tacky as soon as the surface is hydrated. Powdered alginate is slowly poured into the upper part of a vortex created in the water by high-speed stirrer, which must remain submerged to avoid too much aeration. If some clumps do form, the shear should be sufficient to break them up. Dry-mix dispersion can be used when a formulation requires both alginate and other dry ingredients, such as sugars and starches. The dry powders are mixed thoroughly so that the alginate particles are diluted and separated by the other ingredients. This mixture is slowly added to well-stirred water, preferably with a vortex as before, and the other ingredients, often in a ratio of 5 : 1 to 10 : 1, help to keep the alginate particles apart as they are wetted. An even more efficient method of diluting the alginate particles is to use liquid-mix dispersion in which they are wetted with a non-solvent. This can be either a water-miscible nonaqueous liquid, such as ethanol or glycerol, or a water-immiscible liquid, such as vegetable oil. Enough liquid is needed to give a pourable slurry, and this is poured into the water, well agitated as before.

As mentioned before, during the dissolution process, when powders of soluble alginates are wetted with water, the hydration of particles results in each having a tacky surface. Unless some precautions are taken, the particles will rapidly stick together, resulting in clumps that are very slow to completely hydrate and dissolve. Particle size and type affect solubility behavior. Coarse particles are usually preferred because they are easier to disperse and separate, even though they are slower to hydrate and dissolve. Fine particles will dissolve more rapidly, but there is more risk of them clumping together.

Neutral solutions of low- to medium-viscosity alginates can be kept at 25 °C for several years without appreciable viscosity loss, as long as a suitable microbial preservative is added. Solutions of highly polymerized alginates will lose viscosity at room temperature within a year, and to achieve high/stable viscosities, it is better to add calcium ions to a solution of alginate with a moderate DP. All solutions of alginate will depolymerize more rapidly as the temperature is raised. Alginates are most stable in the range of pH = 5–9, while small amounts of calcium greatly increase the stability of sodium alginate solutions [30].

Alginates are insoluble in water-miscible solvents, such as alcohols and ketones. Aqueous solutions of most alginates will tolerate the addition of 10–20% of these solvents, while propylene glycol alginate can tolerate 20–40%. Up to 65% ethanol can be added to triethanolamine alginate without causing precipitation. However, the presence of such solvents in water before dissolving the alginate will hinder

Table 2.5 Highest concentration of organic solvent in a 1% sodium alginate solution.

Organic solvent	Highest concentration (%)	Organic solvent	Highest concentration (%)
Methanol	20	Propanol	10
Ethanol	20	Glycerol	70
Isopropanol	10	Ethylene glycol	70
Butanol	10	Propylene glycol	40

hydration. Table 2.5 shows the highest concentration of organic solvent that can be contained in a 1% sodium alginate solution.

2.4 Viscosity of Alginate Solutions

Many of the uses of alginates depend on their thickening effect, i.e. their ability to increase the viscosity of aqueous systems using relatively low concentrations [11]. Solution viscosity is also important during the production of alginate fibers in the wet-spinning process. The following factors influence the viscosity of alginate solutions.

2.4.1 Effect of Molecular Weight on Solution Viscosity

For polymeric materials, molecular weight has the single most important effect on solution viscosity. Above a certain molecular weight, the effect from separate chains entangling into each other becomes more pronounced with a resultant higher viscosity dependency. Molecular chain stiffness and extension also have an important effect on the intrinsic viscosity as is reflected through the Mark–Houwink–Sakurada equation, which explains how the intrinsic viscosity varies with molecular chain conformation where the exponent generally increases with increasing chain extension. Some measurements have been made on alginate solutions yielding exponent values ranging from 0.73 to 1.31, depending on ionic strength and alginate composition. By increasing the ionic strength, the alginate chain will change from a relatively stiff rod-like conformation to a random coil conformation. A further increase in ionic strength will lead to a further structural change in the chain before it collapses and precipitates.

Although alginate is a natural polymer, manufacturers can control the molecular weight or DP by varying the severity of the extraction conditions to produce products with viscosities in a 1% solution ranging from 10 to 1000 mPa·s, with DP ranging from 100 to 1000 units. Molecular weight is closely related to chain entanglement when alginate is dissolved in water. As the degree of chain entanglement increases with the increase in molecular weight, solution viscosity also increases. As shown

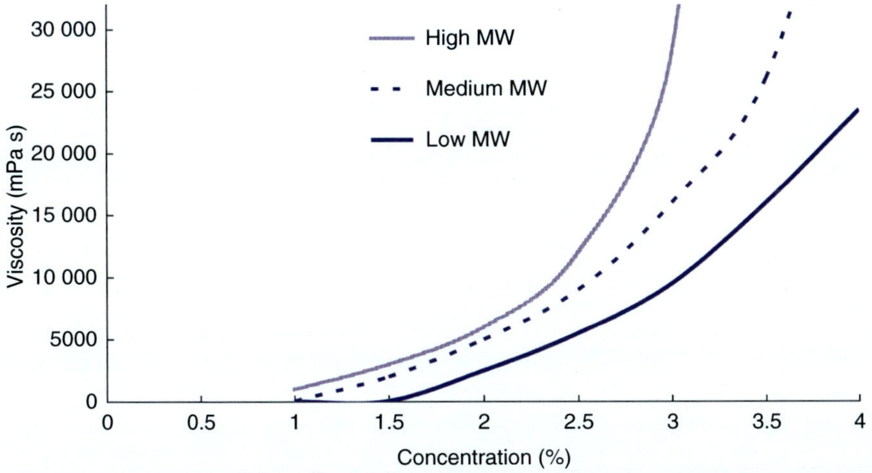

Figure 2.3 Rheological properties of alginate solutions with different molecular weights.

in Figure 2.3, molecular weight has a significant effect on solution viscosity, with high-molecular-weight grade showing extremely high viscosity, which is desired when alginate is used as a thickening agent.

2.4.2 Effect of Concentration on Solution Viscosity

The viscosity of alginate solutions increases sharply as the concentration increases. An empirical equation can be applied, where *a* is a constant related to the DP or molecular weight of the alginate, and *b* is a constant for a particular type of alginate:

The effect of concentration on the rheological properties of alginate solutions is demonstrated in Figure 2.4, where the 0.5% aqueous sodium alginate solution

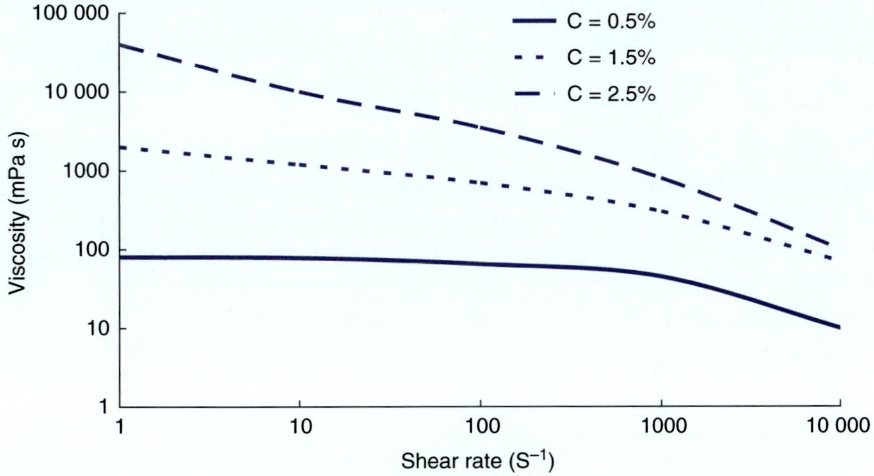

Figure 2.4 Effect of concentration on alginate solution viscosity.

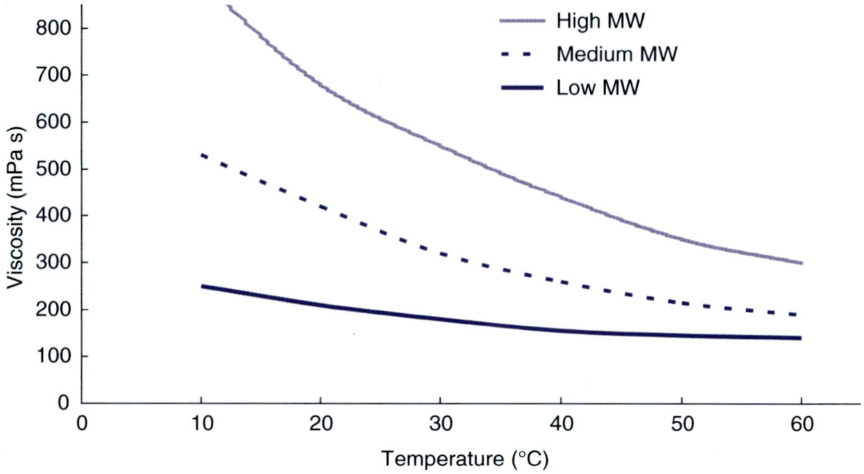

Figure 2.5 Effect of temperature on alginate solution viscosity.

showed near Newtonian behavior, while at higher concentrations, solution viscosities decreased significantly with the increases in shear rates.

2.4.3 Effect of Temperature on Solution Viscosity

The viscosity of alginate solutions decreases as temperature increases, at a rate of about 2.5% per degree Celsius. Since viscosity drops sharply on heating, it is useful to heat a solution during the dissolution process. However, if alginate solutions are kept above 50 °C for several hours, de-polymerization may occur, giving a permanent loss of viscosity and molecular weight. Figure 2.5 shows the effect of temperature on alginate solution viscosity. The rise of temperature significantly affects the chain entanglement, and a significant reduction of solution viscosity is observed as the solution temperature is increased.

2.4.4 Effect of Shear Rate on Solution Viscosity

The viscosities of alginate solutions demonstrate strong shear thinning effects, in particular, solutions with high solid contents show pseudoplastic property in shear rates ranging from 10 to 10 000 S^{-1}. When the solid content drops to 0.5% or below, the alginate solution shows Newtonian behavior when the shear rate is between 1 and 100 S^{-1}. Because alginate has a relatively rigid molecular chain structure, its solution is characterized by shear thinning when solid content and molecular weight are high.

2.4.5 Effect of Salt on Solution Viscosity

As a polymeric salt, sodium alginate shows polyelectrolyte behavior when dissolved in water, with molecular chains showing high degree of extension due to the negative charges from the carboxylic acid groups. Since the molecular chains show relatively extended configuration with high solution viscosity, the addition of salt into the solution can reduce solution viscosity by lowering the electric charges between molecular chains and reducing chain extension. As a result, entanglement between molecular chains is reduced and, correspondingly, there is a drop in solution viscosity with the addition of salt.

2.4.6 Effect of pH on Solution Viscosity

The viscosity of alginate solutions is unaffected over the range of pH 5–11. Below pH 5, the free $-COO^-$ ions in the chain start to become protonated $-COOH$, and as the electrostatic repulsion between chains is reduced, they are able to come closer and form hydrogen bonds, producing higher viscosities. When the pH is further reduced, a gel will form, usually between pH 3 and 4. Above pH 11, slow de-polymerization occurs on storage of alginate solutions, giving a fall in viscosity.

2.5 Polyelectrolyte Properties

Alginate is chemically a polymeric acid that acts as a polyanion at neutral pH and above. The negative charges on the polysaccharide chain result in a notable influence on its overall dimension in solution, which is strongly influenced by the presence of salt. For example, solution viscosity, the radius of gyration, and the persistence length depend on the ionic strength of the medium used. It is common to use aqueous 0.1 M sodium chloride when carrying out physicochemical analyses on alginates. Typically, the intrinsic persistence length of alginate chains is around 12 nm, while the effect of the electrostatic interactions contributes, at an ionic strength of 0.17 M, an additional 3 nm [60].

The chemical composition has a strong influence on the flexibility of the alginate chain, whereby the extension of molecular chain is dependent on its G/M composition, with the intrinsic flexibility of the blocks decreasing in the order MG > MM > GG [49, 56]. In general, alginate with a high content of G-blocks is intrinsically stiffer than alginate with a high M content [54]. It was found that alginate with different compositions displayed a decreasing persistence length upon increasing the M-block and MG-block fractions. In particular, average persistence lengths of 15 and 13 nm were found for alginate from *Laminaria hyperborea* and *Ascophyllum nodosum*, respectively, while for an MG-block-rich alginate, obtained from the acid-soluble fraction, a value as low as 9.5 nm was observed under the same conditions [54].

2.6 The Ion-Exchange Properties of Alginate

As a polymeric acid, alginate can bind divalent metal cations to form hydrogel. This binding ability and affinity to metal ions differ for different metal ions. It was

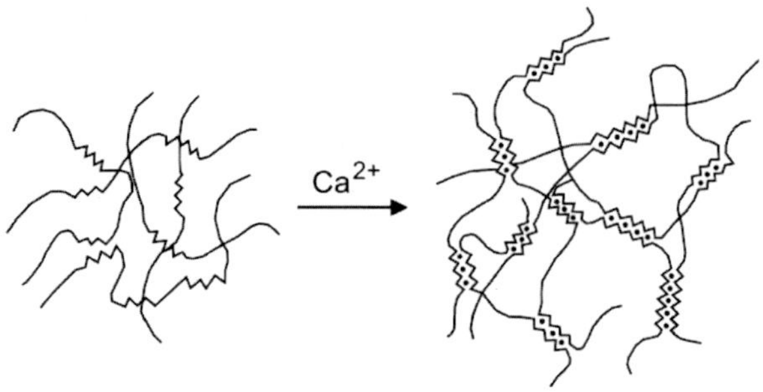

Figure 2.6 An illustration of the "egg box" structure. Source: Adapted with permission from Fig. 13.2, Qin [40].

found that the affinity toward different alkaline-earth divalent ions was in the order $Pb^{2+} > Cu^{2+} > Cd^{2+} > Ba^{2+} > Sr^{2+} > Ca^{2+} > Co^{2+} = Ni^{2+} = Zn^{2+} > Mn^{2+}$ [17, 18]. The molecular structure, in particular the G/M composition, has a strong influence on the affinity of alginic acid to metal ions, with increasing amount of G-blocks favoring the binding of metal ions to alginate [45].

Based on the stereochemical structure of alginic acid, Grant et al. [12] proposed an "egg-box model" to describe the ion-binding ability of alginate. As shown in Figure 2.6, when divalent Ca^{2+} ion is in contact with aqueous sodium alginate solution, it interacts with two adjacent G residues and two G residues in the opposing chain, thus inducing the formation of junction zones, which connect the molecular chains together, leading to the formation of a hydrogel.

It has been found that G-blocks are mainly responsible for this interchain interaction through ionic binding [32, 45], where two facing helical stretches of G sequences bind the divalent ion in a chelate type of binding and there is a geometrical requirement of a cavity formed by the diaxially linked G residues able to accommodate the divalent ions. In contrast, the different configurations of the glycosidic bonds present in the M-blocks and M/G-blocks would not allow such a tight ion entrapment.

The "egg-box model" has been successfully used to explain the ion-exchange and gel-forming properties of alginate, although there have been more studies to give more detailed explanation of the ion-binding nature of alginate [3, 42, 55]. In general, the overall molecular conformation of the alginate molecule is similar to a ridged rod, with the relative stiffness of the polymer chain increasing in the order MG < MM < GG. The affinity of binding and strength of alginate gels are directly related to the polymer composition. Because of this, alginates rich in G-blocks, generally, produce stronger gels than those rich in M-blocks. It is interesting to note that in addition to the formation of ionic binding junction between G-blocks, mixed junctions can also occur between G- and MG-blocks [4].

The binding ability of alginate to different metal ions is characterized as the ion-exchange coefficient, defined as follows:

$$K = \frac{[\text{Divalent metal ions in the gel}][\text{Sodium ions in the solution}]^2}{[\text{Sodium ions in the gel}]^2[\text{Divalent metal ions in the solution}]}$$

Table 2.6 shows the ion-exchange coefficients of different metal ions against sodium alginate. It can be seen that the ion-exchange coefficients of Cu^{2+}, Ba^{2+},

Table 2.6 Ion-exchange coefficients of different metal ions against sodium alginate.

	Source of alginate and M/G ratio	
Metal ions	***L. digitata*** **M/G = 1.60**	***L. hyperborea*** **stipes M/G = 0.45**
$Cu^{2+} - Na^+$	230	340
$Ba^{2+} - Na^+$	21	52
$Ca^{2+} - Na^+$	7.5	20
$Co^{2+} - Na^+$	3.5	4

Table 2.7 Ion-exchange coefficients between Ca^{2+} and Na^+ for alginate extracted from three types of brown seaweeds.

Type of seaweeds	M/G ratio	Ion-exchange coefficient between Ca^{2+} and Na^+
A. nodosum	1.70	7.0
L. digitata	1.60	7.5
L. hyperborea	0.60	20.0

Ca^{2+}, and Co^{2+} with sodium ions are strongly influenced by the M/G ratio of alginate.

Alginates extracted from different species of brown seaweeds have considerably different G, M, GG, MM, and GM contents, and their binding abilities with metal ions also vary greatly. Table 2.7 shows the ion-exchange coefficients between Ca^{2+} and Na^+ for alginate extracted from three types of brown seaweeds.

2.7 Gelling Properties of Alginate

Aqueous sodium alginate solution forms gel when in contact with divalent metal ions, such as Ca^{2+}. During gel formation, calcium ions can be introduced to the polymer in a variety of ways and forms, which produce different types of gels with homogeneous and inhomogeneous structures. For example, when calcium chloride is added to a solution of sodium alginate, the alginate molecules are instantly cross-linked, and through the diffusion of Ca^{2+} ions, an inhomogeneous gel is formed. In another way of gel formation, an internal gelation mechanism can be used, whereby water-insoluble calcium carbonate is first mixed thoroughly in the sodium alginate solution before the calcium carbonate is dissociated by slowly reducing the pH with the addition of glucono-δ-lactone (GDL), which hydrolyzes over a period of 40–60 minutes. This controlled release of the Ca^{2+} is known as internal gelation where the acidity, gelation time, and strength of the resulting gels

Figure 2.7 Alginate-based food gel. Source: Adapted with permission from Fig. 7.5, Qin [40].

can be regulated by carefully adjusting the ratio of $CaCO_3$ to GDL. It is also possible to regulate the gelation process by controlling the particle size of the water-insoluble calcium salt, with smaller particle size favoring a more rapid gelation [6]. Figure 2.7 shows an example of alginate-based food gel formed by internal gelation.

The physical properties of alginate gels are affected by a number of factors, such as the molecular weight of alginate, the M/G ratio, and the concentration of calcium ions. Unlike most other polysaccharides, the gel-forming process of alginate is independent of temperature, with the resultant gel heat stable.

In addition to Ca^{2+}-induced gelation, alginate can form gel through the adjustment of solution pH, since alginic acid is not soluble in water. Once protonated after the addition of acid, alginate precipitates out as an alginic acid gel [5]. The pKa of guluronic acid and mannuronic acid in 0.1 M NaCl is about 3.65 and 3.38, respectively, and a sudden decrease in pH of an alginate solution to below the pKa of the monomers causes the precipitation of alginic acid [19]. When the reduction in pH is performed in a controlled manner, it is possible to form alginic acid gels [5]. This property is illustrated in the anti-heart burn medication Gaviscon, which is an orally administered liquid containing sodium alginate as the active ingredient. When the medicine reaches the stomach, the acidic gastric fluid causes the alginate solution to form an alginic acid gel raft on the surface, which prevents gastric reflux.

2.8 Film-Forming Properties

Alginate has good film-forming properties in that a transparent film can be easily prepared through the evaporation of water from aqueous sodium alginate solution.

If no cross-linking is applied, water-soluble sodium alginate films can be prepared by casting aqueous sodium alginate solution and subsequently drying off the water. It can also be prepared by extrusion of an alginate solution into a non-solvent, such as ethanol or acetone. Pure alginate films are usually too brittle, and in order to improve their softness and flexibility, they can be plasticized with glycerol, sorbitol, or urea. In addition, triethanolamine alginate is used to form soft flexible films.

Water-insoluble alginate films can be made by treating a water-soluble film with a di- or trivalent cation such as Ca^{2+}, or with acid to form alginic acid film which is also not water soluble. A particularly useful way of preparing water-insoluble films is to use zinc alginate, which is soluble in excess ammonia solution. When the NH_3 is evaporated, a water-insoluble zinc alginate film is formed.

2.9 Fiber-Forming Properties

When a solution of sodium alginate is forced through fine holes into a solution of a calcium salt, filaments of calcium alginate are formed. This wet-spinning process has been widely used to make calcium alginate fibers useful in the production of functional wound dressings. Although alginate fibers have limited applications in traditional textile products, due to their high cost and poor stability in alkaline conditions, they are now widely used in wound dressings, showing excellent healing properties. When coming into contact with sodium ions in the body fluids, some of the calcium ions in the fibers are exchanged for sodium ions so that a thin soft gel forms at the interface between the dressing and the wound, which helps to create a moist environment highly beneficial to wound healing.

2.10 Bioactivities of Alginic Materials

Alginate has a number of bioactivities that are useful to promote human health. Through its interaction with cells and tissue, alginate can exhibit enzyme inhibition activities and immune regulation properties that are beneficial to human health.

2.10.1 Enzyme Inhibition Activities of Alginate

Through its binding with zinc and copper ions, alginate can inhibit enzyme activities by depriving them of the vital metal ions. In particular, copper ion is an essential element of tyrosinase, and the application of alginate can impact its activities. Zheng et al. [62] studied the inhibitory effect of sodium alginate on the activity of tyrosinases. Enzyme inhibition kinetics experiments were carried out in vitro and in melanocytes to study the inhibitory effect of sodium alginate on the activity of tyrosinase and melanin deposition. The inhibition mechanism of sodium alginate on tyrosinase was analyzed by fluorescence spectroscopy, and the toxicity of sodium alginate on melanocyte was also investigated. Results showed that within the concentration of 64 mmol/l, the cell viability was above 90%, indicating good

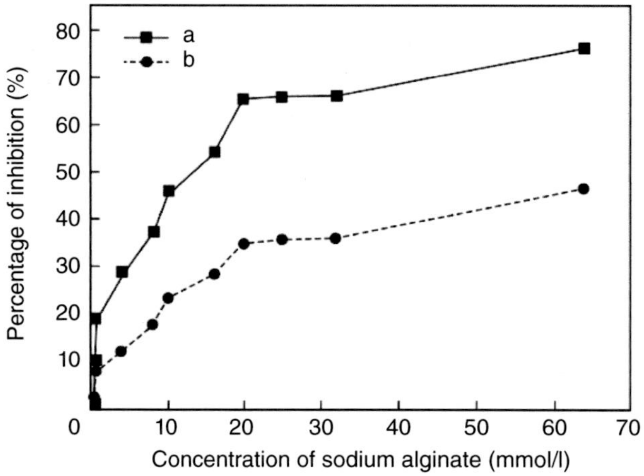

Figure 2.8 Inhibition effect of sodium alginate against tyrosinase in solution (a) and within melanocyte cells (b).

biocompatibility. Sodium alginate had a certain inhibitory effect on tyrosinase activity and melanin production, which was in a concentration-dependent manner below the concentration of 20 mmol/l. The endogenous fluorescence of tyrosinase was partly quenched after interacting with sodium alginate. The shift of the maximum emission wavelength appearing in the fluorescence spectrum of tyrosinase indicated that the interaction between sodium alginate and tyrosinase changed the hydrophobic environment of its amino acid residues. Figure 2.8 shows the inhibition effect of sodium alginate against tyrosinase in solution and within melanocyte cells.

2.10.2 Biocompatibility and Cell Activities of Alginate

As a natural polymer, alginate has good biocompatibility and has been widely used in the food and pharmaceutical industries. It was included in the US and British Pharmacopeia in 1938 and 1963, respectively. In a study by Blair et al. [2], it was shown that when calcium alginate fibers were implanted into mesenteric vein, they did not cause intestine obstruction, although there was some deposition of calcium on the injured site. In another study, Lansdown and Payne [26] implanted calcium alginate fibers in the subcutaneous tissue of a rat, and after 24 hours, 7 days, 28 days, and 12 weeks, test results showed that these fiber showed no sign of degradation. Although there was a slight foreign body reaction, the implanted fibers were gradually covered by a vascularized membrane containing fibroblasts, indicating the fact that these fibers were nontoxic to the rat. Other studies also confirmed that alginate has good biocompatibility [57].

In addition to its good biocompatibility, as a natural component of plant cell walls, alginate has special affinity with various cells and can affect cell activities [38]. Skjak-Braek and Espevik [43] found that the M-block in alginate is similar to lipopolysaccharide (LPS) and has a binding site with macrophages, which can produce chemotaxis to monocytes through interaction with membrane proteins. Otterlei et al. [34, 35] studied the effect of alginate in stimulating human monocytes to produce tumor necrosis factor-alpha (TNF-alpha), interleukin-1, and interleukin-6. Results showed that the cell-stimulating effect for the high M alginate is about 10 times that for high G alginate, indicating the M-blocks being mainly responsible for the cell activities of alginate. Skjak-Braek et al. [43] showed that when the M-blocks were enzymatically converted to G-blocks, alginate lost its activity in the induction of tumor necrosis factor-alpha. Zimmerman et al. [25, 63] compared the mitotic activities of several samples of alginate and found that after purification through electrophoresis and dialysis, the purified alginate showed no mitotic activities, which may indicate that the cell activities are mainly from low-molecular-weight alginate oligosaccharides.

2.11 Summary

Alginate is a water-soluble natural polymer, commonly used in the form of sodium alginate. In its various fields of applications, alginate can be made into a variety of end products with different chemical and physical structures, with the resultant materials possessing a large number of unique chemical, physical, and biological properties.

References

1 Atsuki, K. and Tomoda, Y. (1926). Studies on seaweeds of Japan, I. The chemical constituents of *Laminaria*. *J. Soc. Chem. Ind. Jpn.* 29: 509–517.

2 Blair, S.D., Backhouse, C.M., Harper, R. et al. (1988). Comparison of absorbable materials for surgical haemostatis. *Br. J. Surg.* 75 (10): 969–971.

3 Braccini, I. and Pérez, S. (2001). Molecular basis of Ca^{2+}-induced gelation in alginates and pectins: the egg-box model revisited. *Biomacromolecules* 2: 1089–1096.

4 Donati, I., Holtan, S., Mørch, Y.A. et al. (2005). New hypothesis on the role of alternating sequences in calcium-alginate gels. *Biomacromolecules* 6: 1031–1040.

5 Draget, K.I., Skjåk Bræk, G., and Smidsrød, O. (1994). Alginic acid gels: the effect of alginate chemical composition and molecular weight. *Carbohydr. Polym.* 25: 31–38.

6 Draget, K.I., Strand, B., Hartmann, M. et al. (2000). Ionic and acid gel formation of epimerised alginates, the effect of AlgE4. *Int. J. Biol. Macromol.* 27: 117–122.

7 Ertesvåg, H., Doseth, B., Larsen, B. et al. (1994). Cloning and expression of an *Azotobacter vinelandii* mannuronan C-5-epimerase gene. *J. Bacteriol.* 176: 2846–2853.

8 Ertesvåg, H., Valla, S., and Skjåk-Bræk, G. (1996). Genetics and biosynthesis of alginates. *Carbohydr. Eur.* 14: 14–18.

9 Fisher, F.G. and Dörfel, H. (1955). Die Polyuronsäuren der Braunalgen (Kohlenhydrate der Algen). *Z. Physiol. Chem.* 302: 186–203.

10 Gacesa, P. (1992). Enzymic degradation of alginates. *Int J Biochem* 24: 545–552.

11 Glicksman, M. (1969). *Gum Technology in the Food Industry*. New York: Academic Press.

12 Grant, G.T., Morris, E.R., Rees, D.A. et al. (1973). Biological interactions between polysaccharides and divalent cations: the egg-box model. *FEBS Lett.* 32: 195–198.

13 Grasdalen, H., Larsen, B., and Smidsrød, O. (1979). A p.m.r. study of the composition and sequence of uronate residues in alginates. *Carbohydr. Res.* 68: 23–31.

14 Grasdalen, H., Larsen, B., and Smisrød, O. (1981). ^{13}C-n.m.r. studies of monomeric composition and sequence in alginate. *Carbohydr. Res.* 89: 179–191.

15 Hartmann, M., Holm, O.B., Johansen, G.A.B. et al. (2002). Mode of action of recombinant *Azotobacter vinelandii* mannuronan C-5 epimerases AlgE2 and AlgE4. *Biopolymers* 63: 77–88.

16 Hartmann, M., Duun, A.S., Markussen, S. et al. (2002). Time resolved ^{1}H and ^{13}C NMR spectroscopy for detailed analyses of the *Azotobacter vinelandii* mannuronan C-5 epimerase reaction. *Biochim. Biophys. Acta, Gen. Subj.* 1570: 104–112.

17 Haug, A. and Smidsrød, O. (1967). Strontium-calcium selectivity of alginates. *Nature* 215: 757.

18 Haug, A. and Smidsrød, O. (1970). Selectivity of some anionic polymers for divalent metal ions. *Acta Chem. Scand.* 24: 843–854.

19 Haug, A., Larsen, B., and Smidsrød, O. (1963). The degradation of alginates at different pH values. *Acta Chem. Scand.* 17: 1466–1468.

20 Haug, A., Larsen, B., and Smidsrød, O. (1966). A study of the constitution of alginic acid by partial hydrolysis. *Acta Chem. Scand.* 20: 183–190.

21 Haug, A., Larsen, B., and Smidsrød, O. (1966). A study on the constitution of alginic acid by partial hydrolysis. *Acta Chem. Scand.* 20: 183–190.

22 Haug, A., Larsen, B., and Smidsrød, O. (1967). Studies on the sequence of uronic acid residues in alginic acid. *Acta Chem. Scand.* 21: 691–704.

23 Høidal, H.K., Ertesvåg, H., Skjåk-Bræk, G. et al. (1999). The recombinant *Azotobacter vinelandii* mannuronan C-5-epimerase AlgE4 epimerizes alginate by a nonrandom attack mechanism. *J. Biol. Chem.* 274: 12316–12322.

24 Holtan, S., Bruheim, P., and Skjåk-Bræk, G. (2006). Mode of action and subsite studies of the guluronan block-forming mannuronan C-5 epimerases AlgE1 and AlgE6. *Biochem. J.* 395: 319–329.

25 Klock, G., Frank, H., Houben, R. et al. (1994). Production of purified alginates suitable for use in immunoisolated transplantation. *Appl. Microbiol. Biotechnol.* 40 (5): 638–643.

26 Lansdown, A.B. and Payne, M.J. (1994). An evaluation of the local reaction and biodegradation of calcium sodium alginate (Kaltostat) following subcutaneous implantation in the rat. *J. R. Coll. Surg. Edinb.* 39 (5): 284–288.

27 Larsen, B. and Painter, T.J. (1969). The periodate-oxidation limit of alginate. *Carbohydr. Res.* 10: 186–187.

28 Larsen, B., Smidsrød, O., Painter, T.J. et al. (1970). Calculation of the nearest neighbour frequencies in fragments of alginate from the yields of free monomers after partial hydrolysis. *Acta Chem. Scand.* 24: 726–728.

29 Leo, W.J., McLoughlin, A.J., and Malone, D.M. (1990). Effects of sterilization treatments on some properties of alginate solutions and gels. *Biotechnol. Prog.* 6: 51–53.

30 McNeely, W.H. and Pettitt, D.J. (1973). Algin. In: *Industrial Gums*, 2nde (ed. R.L. Whistler), 52–78. New York: Academic Press.

31 Mørch, Y.A., Donati, I., Strand, B.L. et al. (2007). Molecular engineering as an approach to design new functional properties of alginate. *Biomacromolecules* 8: 2809–2814.

32 Morris, E.R., Rees, D.A., Thom, D. et al. (1978). Chiroptical and stoichiometric evidence of a specific, primary dimerisation process in alginate gelation. *Carbohydr. Res.* 66: 145–154.

33 Morris, E.R., Rees, D.A., and Thom, D. (1980). Characterization of alginate composition and block structure by circular dichroism. *Carbohydr. Res.* 81: 305–314.

34 Otterlei, M., Ostgaard, K., and Skjak-Braek, G. (1991). Induction of cytokine production from human monocytes stimulated with alginate. *J. Immunother.* 10: 286–291.

35 Otterlei, M., Sundan, A., Skjak-Braek, G. et al. (1993). Similar mechanisms of action of defined polysaccharides and lipopolysaccharides: characterization of binding and tumor necrosis factor alpha induction. *Infect. Immun.* 61 (5): 1917–1925.

36 Painter, T.J., Smidsrød, O., and Haug, A. (1968). A computer study of the changes in composition-distribution occurring during random depolymerisation of a binary linear heteropolysaccharide. *Acta Chem. Scand.* 22: 1637–1648.

37 Penman, A. and Sanderson, G.R. (1972). A method for the determination of uronic acid sequence in alginates. *Carbohydr. Res.* 25: 273–282.

38 Pueyo, M.E., Darquy, S., Capron, F. et al. (1993). *In vitro* activation of human macrophages by alginate-polylysine microcapsules. *J. Biomater. Sci. Polym. Ed.* 5 (3): 197–203.

39 Qin, Y. (2016). *Medical Textile Materials*. Kidlington: Woodhead Publishing.

40 Qin, Y. (2018). *Bioactive Seaweeds for Food Applications*. San Diego, CA: Academic Press.

41 Schimdt, E. and Vocke, F. (1926). Zur Kenntnis der Polyglykuronsäuren. *Chem. Ber.* 59: 1585–1588.

42 Sikorski, P., Mo, F., Skjåk-Bræk, G. et al. (2007). Evidence for egg-box-compatible interactions in calcium-alginate gels from fiber X-ray diffraction. *Biomacromolecules* 8: 2098–2103.

43 Skjak-Braek, G. and Espevik, T. (1996). Application of alginate gels in biotechnology and biomedicine. *Carbohydr. Eur.* 14: 19–25.

44 Skjåk-Bræk, G., Smidsrød, O., and Larsen, B. (1986). Tailoring of alginates by enzymatic modification in vitro. *Int. J. Biol. Macromol.* 8: 330–336.

45 Smidsrød, O. (1974). Molecular basis of some physical properties of alginates in the gel state. *Faraday Discuss. Chem. Soc.* 57: 263–274.

46 Smidsrød, O. and Haug, A. (1972). Properties of poly-(1,4-hexuronates) in the gel state. II. Comparison of gels of different chemical composition. *Acta Chem. Scand.* 26: 79–88.

47 Smidsrød, O. and Painter, T. (1973). Effect of periodate oxidation upon the stiffness of the alginate molecule in solution. *Carbohydr. Res.* 26: 125–132.

48 Smidsrod, O. and Whittington, S.G. (1969). Monte Carlo investigation of chemical inhomogeneity in polymers. *Macromolecules* 2: 42–44.

49 Smidsrød, O. and Whittington, S.G. (1969). Monte Carlo investigation of chemical inhomogeneity in copolymers. *Macromolecules* 2: 42–44.

50 Smidsrød, O., Haug, A., and Larsen, B. (1963). Degradation of alginate in the presence of reducing compounds. *Acta Chem. Scand.* 17: 2628–2637.

51 Smidsrød, O., Haug, A., and Larsen, B. (1963). The influence of reducing substances on the rate of degradation of alginates. *Acta Chem. Scand.* 17: 1473–1474.

52 Smidsrød, O., Haug, A., and Larsen, B. (1966). The influence of pH on the rate of hydrolysis of acidic polysaccharides. *Acta Chem. Scand.* 20: 1026–1034.

53 Smidsrød, O., Haug, A., and Larsen, B. (1967). Oxidative-reductive depolymerization: a note on the comparison of degradation rates of different polymers by viscosity measurements. *Carbohydr. Res.* 5: 482–485.

54 Smidsrød, O., Glover, R.M., and Whittington, S.G. (1973). The relative extension of alginates having different chemical composition. *Carbohydr. Res.* 27: 107–118.

55 Steginsky, C.A., Beale, J.M., Floss, H.G. et al. (1992). Structural determination of alginic acid and the effects of calcium binding as determined by high-field n.m.r. *Carbohydr. Res.* 225: 11–26.

56 Stokke, B.T., Smidsrød, O., and Brant, D.A. (1993). Predicted influence of monomer sequence distribution and acetylation on the extension of naturally occurring alginates. *Carbohydr. Polym.* 22: 57–66.

57 Suzuki, Y., Nishimura, Y., Tanihara, M. et al. (1998). Evaluation of a novel alginate gel dressing: cytotoxicity to fibroblasts in vitro and foreign-body reaction in pig skin in vivo. *J. Biomed. Mater. Res.* 39 (2): 317–322.

58 Svanem, B.I.G., Skjåk-Bræk, G., Ertesvåg, H. et al. (1999). Cloning and expression of three new *Azotobacter vinelandii* genes closely related to a previously described gene family encoding mannuronan C-5-epimerases. *J. Bacteriol.* 181: 68–77.

59 Timell, T.E. (1964). The acid hydrolysis of glycosides: I. General conditions and the effect of the nature of the aglycone. *Can. J. Chem.* 42: 1456.

60 Vold, I.M.N., Kristiansen, K.A., and Christensen, B.E. (2006). A study of the chain stiffness and extension of alginates, in vitro epimerized alginates, and periodate-oxidized alginates using size-exclusion chromatography combined with light scattering and viscosity detectors. *Biomacromolecules* 7: 2136–2146.

61 Wong, T.Y., Preston, L.A., and Schiller, N.L. (2000). Alginate lyase: review of major sources and enzyme characteristics, structure-function analysis, biological roles, and applications. *Annu. Rev. Microbiol.* 54: 289–340.

62 Zheng, X., Chen, Z., Zhang, D. et al. (2019). The inhibition of sodium alginate against tyrosinase. *CSDC* 49 (6): 388–392.

63 Zimmermann, U., Klock, G., Federlin, K. et al. (1992). Production of mitogen-contamination free alginates with variable ratios of mannuronic acid to guluronic acid by free flow electrophoresis. *Electrophoresis* 13: 269–274.

3

Industrial Applications of Alginic Materials

3.1 Introduction

Alginate has a unique chemical structure that enables it to gel at low temperature, with the resultant gel possessing good heat stability. This novel gelling property makes it ideal for use as thickener, stabilizer, and restructuring agent, as well as for encapsulating active enzymes and live bacteria, and for protective coating of fruits and vegetables. As a food ingredient, the alginate intake in the diet has a number of physiological effects. Because of the presence of acid in the stomach, alginate can gel in situ, thereby increasing satiety and reducing absorption and digestion rates of other macronutrients in the diet [16, 20, 43]. In addition to its main application in the food industry, since its discovery in 1881, alginate has become a vastly utilized polymer with a large variety of applications. In particular, the novel properties of alginate have found applications in the paper, textile, agricultural, pharmaceutical, and biomedical industries.

3.2 Functional Properties of Alginic Material

Alginate has many unique properties, some of them are described below.

3.2.1 Alginate as a Thickening Agent

Alginate is a polyelectrolyte and when dissolved in water, its molecular chains are not completely flexible, forming highly viscous solutions that exhibit strong shear-thinning effects. This gives alginate solution very high viscosity during storage, while the reduced viscosity upon processing helps to assist material flow. This novel thickening effect is highly useful in the food industry, and alginate is widely used in ice creams, desserts and savory sauces, jams, marmalades, and fruit sauces. Alginate can also be used in conjunction with other thickening agents to improve the acceptability of a number of low-fat processed foods.

Alginate Fibers and Wound Dressings: Seaweed Derived Natural Therapy, First Edition. Yimin Qin.

Figure 3.1 An illustration of the various applications of alginate-based food gel.

3.2.2 Alginate as a Gelling Agent

Alginate can form thermo-irreversible and heat stable gel when sodium alginate is in contact with calcium or other multivalent metal ions. The room temperature gelling function is particularly valuable in the restructuring of foods that may become damaged or oxidized under high temperatures. Alginate gel is commonly used in reconstituted onion rings, in which it enables the production of products of regular size and consistency. The gel-forming property can also be applied in meat, seafood, fruit, vegetable, and some extruded food products. The ability of alginate to form synergistic gels with other polysaccharides such as pectin or chitosan provides further scope for its use in food. In particular, alginate–pectin synergistic gels are used in jams, jellies, and fruit fillings. This type of gel is thermo-reversible and gives a higher viscosity than either individual component. Figure 3.1 shows the various applications of alginate-based food gel.

3.2.3 Alginate as a Film-Forming Agent

As a natural polymer, alginate can form film when its aqueous solution is cast as a thin layer and subsequently dried by removing the water through evaporation. This property can be used in edible films and coatings for foodstuffs, which have a high potential owing to the current call for reduction or replacement of nonbiodegradable or nonrecyclable food packaging, especially when sodium alginate food films show good tensile strength, flexibility, and resistance to tearing and are impermeable to oils. In addition, antimicrobial agents can be incorporated in the alginate film to provide an effective barrier to microbial surface spoilage of vegetables, meat, and fish products. Alginate films can be used to protect frozen fish from oxidation and loss of water by stabilizing the ice layer and making it more impermeable to oxygen and moisture. Meat carcasses and meat pieces can be protected by a calcium alginate film, which both reduces water loss and improves food safety.

3.2.4 Alginate as a Stabilizer

Alginate forms stable gels both at high or low temperatures and at low pH. In the food processing industry, the gelling property of alginate can be used as food stabilizers. Ice cream was the first application of alginate as a stabilizer, where alginate allows control of the ice cream's viscosity, increases heat-shock resistance, reduces shrinkage and ice crystal formation, and enables ice cream to have a prolonged meltdown, which is a desired characteristic. In order to avoid precipitation, sodium alginate is usually mixed with sodium phosphate, which can bind excess calcium ions in the mixture to prevent the formation of calcium alginate.

3.2.5 Alginate for Encapsulation and Immobilization

The gel-forming property of alginate can be used to encapsulate bioactive agents in food processing and other industries, where the bioactive agents are first dispersed in a sodium alginate solution before being precipitated upon gelation. This technology is particularly useful to protect those ingredients that are unstable, volatile, or particularly reactive. Alginate-based encapsulation renders a cover around these ingredients, thereby providing stability and protection for the whole product.

Figure 3.2 shows an illustration of alginate-based beads which are commonly used in immobilization technologies for food processing and biotechnology to entrap cells or enzymes. Alginate-based immobilization technology can be used to produce a wide range of bacterial metabolites, including enzymes, amino acids, organic acids, and alcohols. For example, alginate immobilization technology can be used in the production of fermented beverages to entrap and reuse various cell yeast cultures or in the dairy industry, where various lactic acid-producing bacteria are used as a starter culture.

Figure 3.2 An illustration of alginate-based beads.

3.3 Industrial Applications of Alginate

The industrial applications of alginate are mainly based on its thickening, gelling, and film-forming properties. Some of the applications are summarized below.

3.3.1 Food Ingredients

Alginate has long been used as a food ingredient, where its thickening properties are useful in sauces, syrups, toppings for ice cream, etc., while its ability to form gel is useful in instant milk desserts and jellies, bakery filling cream, fruit pies, animal foods, and reformed fruit. Utilizing its stabilizing and emulsifying properties, sodium alginate is also added to ice creams to break the ice crystals [12, 21, 31, 38, 40, 44].

In 1986, the US Department of Agriculture approved the use of alginate as a binder in restructured foods, where pieces of flaked, sectioned, or chunked meat are bound by the alginate gel to resemble intact cuts of meat, which can be served as nuggets, roasts, loaves, and steaks. During the production process, a mixed powder of sodium alginate, calcium carbonate, lactic acid, and calcium lactate is mixed with the raw meat before the resultant calcium alginate gel binds the meat together. This binding mixture can be used to replace the sodium chloride and phosphate salts commonly used, thereby reducing the sodium level in the restructured products [2, 46, 71].

The gelling property of alginate is also useful in the production of re-formed foods. For example, shrimp or crab meat analog products can be produced by using alginate and proteins such as soy protein concentrate or sodium caseinate [47]. During the production process, a mixture of the two ingredients is extruded into an aqueous calcium chloride bath to form edible fibers which are then frozen, thawed, chopped, coated with sodium alginate, and formed in an appropriately shaped mold. After further freezing and thawing, a product analogous to natural shrimp is obtained. Similar processes have been used to produce meat substitutes [76], structured fruit products, and structured potato products such as croquettes and French fries [3, 13, 53].

As a natural food ingredient, alginate can be added to sauces, syrups, and ice cream toppings to increase viscosity and improve texture. Alginate can also be used to thicken pie fillings whereby softening of the pastry by liquid from the filling is reduced. Addition of alginate can make icings nonsticky and allow the baked goods to be covered with plastic wrap. When thickened with alginate, water-in-oil emulsions such as mayonnaise and salad dressings are less likely to separate into their original oil and water phases. In chocolate milk, the cocoa can be kept in suspension by an alginate/phosphate mixture, although in this application it faces strong competition from carrageenan. The addition of small amounts of alginate can thicken and stabilize whipped cream.

Alginate can also be used to make edible dessert jellies. Unlike other food gels such as those made from gelatin and agar, the alginate-based gels can be made at room temperature, with the resultant gel heat stable since it possesses a stable structure cross-linked with divalent metal ions. Because they do not melt, alginate jellies have a different, firmer mouth feel when compared to gelatin jellies, which can be made to soften and melt at body temperature. During the production process, mixtures of calcium salts and sodium alginate can be made to set to a gel at different rates, depending on the rate at which the calcium salt dissolves. Gel formation can also be delayed even after everything is mixed together, which is done using a gel-retarder that reacts with the calcium before the alginate does, so no calcium is available to the alginate until all the retarder is used. In this way, gel formation can be delayed for several minutes if desired, such as when other ingredients need to be added and mixed before the gel starts to set. Figure 3.3 shows an illustration of alginate-based food gels, which offer a tasty mouth feel when mixed with a proper sauce, even though it contains very little energy.

The film-forming property of alginate can be used to help preserve frozen fish. The oils in oily fish such as herring and mackerel can become rancid through oxidation even when they are quickly frozen and stored at low temperatures. If it is frozen in a calcium alginate jelly, the fish is protected from the air and rancidity from oxidation is very limited. The jelly thaws with the fish so they are easily separated. If beef cuts are coated with calcium alginate films before freezing, the meat juices released during thaw are re-absorbed into the meat and the coating also helps to protect the meat from bacterial contamination. The calcium alginate coating can be removed by re-dissolving it with sodium polyphosphate if necessary.

Alginate films are also used in sausage casing or sausage skin. In the past, sausage was produced using naturally available casings such as intestines of pigs, sheep, goats, cattle, and sometimes horses. Because of the shortage of natural

(2) Spiced alginate gel

(1) Noodle like alginate gel

(3) Shredded alginate gel fried with rice

Figure 3.3 An illustration of alginate-based food gels.

sources, nowadays sausage casings are mainly made from film-forming polymers, utilizing collagen, cellulose, and plastics such as polyamide, polypropylene, and polyethylene. As a film-forming natural polymer, alginate can provide a layer of protective film on the sausage through its gel formation with calcium ions. During the production process, meat mixture is extruded to form the core and a layer of sodium alginate is first coated before being subjected to an aqueous calcium chloride solution to induce gelation. A layer of calcium alginate film is then formed on the sausage which is heat stable and can offer a strong casing for the sausage. It has the required strength and flexibility and can protect the sausage against water and oil losses.

3.3.2 Medical and Pharmaceutical Uses

Alginate has many applications in the pharmaceutical industry. For example, alginic acid is used as disintegrating agent in tablets because it is insoluble in water but can swell when placed in water. Although it is more expensive than traditional disintegrating agents such as starch, the mechanical strength of the final tablet is greater than that of starch-based tablets. Sodium alginate is used in some liquid medicines to increase viscosity and improve the suspension of solids. Capsules containing sodium alginate and calcium carbonate are used to protect inflamed areas near the entrance to the stomach. The acidity of the stomach causes the formation of insoluble alginic acid and carbon dioxide. As a result, alginic acid rises to the top of the stomach contents and forms a protective layer.

Some of the many medical and pharmaceutical applications of alginate are summarized below.

3.3.2.1 Dental Impression

Based on its ability to form cold-setting gels, alginate has long been used in dental impression [55, 70]. As shown in Table 3.1, a typical dental impression formulation consists of sodium alginate, calcium sulfate, sodium tripolyphosphate, and diatomaceous earth. During application, water is added into the mixed powder before being placed in the mouth. Upon the slow release of calcium ions, the impression material solidifies before being taken out and placed in 10–20% $MgSO_4$ solution for 10–15 minutes.

Table 3.1 Composition of a dental impression formulation.

Ingredient	Amount (%)
Sodium alginate	15
Calcium sulfate dihydrate	15
Sodium tripolyphosphate	2
Diatomaceous earth	68

3.3.2.2 Therapeutic Cell Entrapment

Calcium alginate gel can be used to entrap living cells for therapeutic treatment of human illness. The entrapment in Ca-alginate spheres is a well-established process that can be carried out in a single step under very mild conditions, where a cell suspension is mixed with sodium alginate solution before the mixture is dripped into a solution containing Ca^{2+} and the droplets instantaneously form gel-spheres entrapping the cells in a three-dimensional lattice. In addition to cell entrapment, alginate-immobilized cells can be used in cell transplantation, where the main purpose of the gel is to act as a barrier between the transplant and the immune system of the host. In this way, parathyroid cells can be used for the treatment of hypocalcemia and dopamine-producing adrenal chromaffin cells can be used for the treatment of Parkinson's disease. The alginate-based cell entrapment technique can also be used for insulin-producing cells that are useful in the treatment of Type I diabetes.

In cell immobilization and transplantation, the alginate-based gel should possess high mechanical and chemical stability, controllable swelling properties, low content of toxic, pyrogenic, and immunogenic contaminants. They should also have defined pore size and a narrow pore size distribution. These performance requirements are met by using purified alginates, controlling gelling ions, and gelling kinetics. In addition, alginate can be used in combination with other polymers to achieve required product characteristics, for example, by forming polyelectrolyte complex with chitosan. The mechanical and swelling properties of the gel beads depend strongly on the monomeric composition, block structure, and molecular weight of the alginate, where high mechanical stability can be achieved by using alginate with a high content of guluronic acid.

3.3.2.3 Controlled Release of Drugs

Alginate is used in the controlled release of medicinal drugs, in particular for the release of drugs in the intestine [4]. Since alginic acid is water insoluble, drugs embedded in an alginate matrix can withstand the acid environment of the stomach and once into the intestine, they are released upon the dissolution of alginate in the slightly alkaline condition. The controlled release properties of alginate microspheres can be further improved by coating with chitosan to improve mechanical strength and release profile.

3.3.2.4 Alginate Oligoelectrolytes as a Mucin Polymer Network Modifier

Alginate oligomers are able to modify both the gelling kinetics and the apparent equilibrium properties of alginate gels. It has been shown that these oligomers are able to transiently modify mucin network structures, which may provide a basis for the treatment of pathological respiratory conditions as well as for a general manipulation of mucosal surfaces for drug delivery systems [49, 50, 54].

3.3.2.5 Oligoguluronates as Modifiers of Cystic Fibrosis Mucus

In patients diagnosed with cystic fibrosis, the most common lethal genetic disease among Caucasians, the general pathological picture is described by a thick intractable mucus in multiple organs of which the lungs are the most problematic, accounting for well over 90% of the mortality [56]. Mucus rheology plays a major role in the mucociliary clearance of the lungs, but in the case of cystic fibrosis, the highly viscous mucus combined with a hyperinflammatory state leads to malfunctioning of the mucociliary clearance. In addition to the normal mucins, a number of non-mucin macromolecular components, such as DNA, actin, and bacterial polysaccharides, appear at high levels in CF mucus and are suspected to make a substantial contribution to the viscoelastic properties. For example, high molecular weight alginate such as produced by *Pseudomonas aeruginosa*, a most common inflammatory microorganism in the CF lung, interacts with mucin in a synergistic manner and this interaction bears the fingerprint of being of an electrostatic nature. The possibility of applying low molecular weight oligoelectrolytes to normalize the rheology of mucus by a shielding off of the interaction sites between mucin and other macromolecular components therefore became pertinent. The general idea behind this approach was that these types of interactions, which lead to increased mucus mechanical properties, could be eliminated through electrostatic competitive inhibition by charged oligomers too small to create intermolecular cross-links. G-blocks, typically with a degree of polymerization of 10–20, were chosen as an appropriate candidate for this purpose as they are known to be non-immunogenic [5, 6, 15, 19, 23, 58, 77, 86].

3.3.3 Wound Dressings and Hemostatic Agent

Alginate can be made into fibers through a wet spinning process with the fibers further processed into nonwoven wound dressings [1, 62, 65, 81–83]. Alginate-based "biopaper" was also developed for use as bandages and similar medical uses where the hemostatic properties of alginates are useful. Bioactive papers were also made from staple fibers in which enzymes were entrapped [35–37]. During contact with body fluid and its metal ions, the unique polysaccharide structure of alginate plays an important role in the high absorbency and moisture retention, hemostasis, wound healing promotion, antimicrobial, detoxication, cosmetic and other bioactivities, making alginate fibers widely applied in medical and hygiene products such as medical dressings, face masks, female hygiene products, baby diapers, and adult incontinence products [59–61, 63].

3.3.4 Immobilization of Biocatalysts

Alginate can be used to immobilize biocatalysts such as enzymes and whole cells. In the biotechnology industry, enzymes are used to convert glucose to fructose, to produce L-amino acids for use in foodstuffs, and to synthesize new penicillins after hydrolysis of penicillin G. Similarly, whole cells are used to promote the conversion of starch to ethanol for beer brewing, and for the continuous production of yogurt. In these processes, immobilization of enzymes and cells by fixing them to the surface of an insoluble solid or entrapping them in a polymeric material is a practical way to realize the potential of biocatalyst-assisted processes.

In the 1970s, many single enzymes were isolated, immobilized, and used. More recently, it has been found that it is easier, more economical, and often more effective to immobilize whole cells, which contain multi-enzyme systems. An added advantage of immobilizing cells is the increased stability, with a half-life of one day for ordinary suspended cells increasing to 30 days for immobilized or resting cells [84]. Alginate gels are particularly suited for entrapping biocatalysts such as cells and enzymes. During the preparation process, the cell suspension is usually mixed with 2–4% sodium alginate solution, which is then extruded as drops into aqueous calcium chloride solution with a concentration of 0.05–0.1 M. An immediate skin forms around the drop and as calcium ions gradually diffuse inwards, a gel forms. The size of the beads can be regulated from the size of the needle or nozzle, usually 0.2–1.0 mm but up to 5 mm. The fresh beads can be separated and used or they can be dried, which increases their strength and reduces their ability to swell so they contain, when rewetted, more cells per unit volume. High gel strength is required for good immobilization and therefore high G alginate is more suitable in this application [14, 32, 33, 67, 78].

Johansen and Flink [26–29] used the internal gelation principles for food gels to immobilize yeast cells, where sodium alginate, an insoluble calcium salt, and D-glucono-1,5-lactone are mixed in water, and as the lactone slowly hydrolyzes, the solution pH is lowered to enable the release of calcium ions to initiate gelation gradually from within the solution. The resulting immobilizates have particles of higher strength, with at least equal fermentation rates, when compared to externally gelled material. Rochefort et al. [68] stabilized calcium alginate gels by washing with 0.1 M aluminum nitrate. Burns et al. [7] produced dried spheres of calcium alginate containing magnetite and found they have good potential as a support for enzyme immobilization.

Cell immobilization with alginate can be done under mild conditions with little loss of activity of the cells and the activity is often stable for extended periods of time. Temperatures can be 0–100 °C and the pH neutral but any buffers used must not contain citrate or phosphate, which can remove calcium ions from the gel and can lead to its breakdown. On the other hand, the cells can be recovered if necessary by adding a sequestering agent for the calcium ions such as polyphosphate or EDTA. Once the calcium ions are removed from the gel, its structure is lost and it changes to a liquid with the cells suspended in it.

Alginate has been used for cell or enzyme immobilization in a large number of processes, some of them are listed below:

- Production of ethanol from starch [45];
- Beer brewing with immobilized yeast [52];
- Production of citric acid [39];
- Continuous yogurt production [57];
- Fermentation to produce butanol and isopropanol [72];
- Continuous acetone-butanol production [18];
- Pilot-plant production of prednisolone from hydrocortisone [34]; and
- Glycerol production from the marine alga, *Dunaliella tertiolecta* [22].

3.3.5 Controlled Release of Active Agents

Similar to the immobilization of enzyme and cells, bioactive agents such as herbicides and bio-stimulants can also be combined with alginate to achieve controlled release. By incorporating air into the beads, they can also be made to float [10, 11, 17].

3.3.6 Textile Printing Paste

One of the main applications of alginate is as a thickening agent in textile printing, where the properties of alginate are outstanding among all other competing materials such as starch, carboxymethyl cellulose, and polyvinyl alcohol. The alginate-based printing paste is unique in that in addition to its excellent viscoelastic properties, the alginate molecules in the paste do not react with reactive dyes that are now commonly used in textile printing. Many of the standard thickeners, such as starch, react with these dyes, leading to lower color yields and sometimes insoluble products, which are not easily washed out, resulting in a fabric with poor handle. In its hydrated state, the alginate molecule is surrounded by negatively charged carboxylic acid groups and there is minimal reaction between alginate and reactive dyes. In addition, sodium alginate can be easily washed out of the finished textile, making it the best thickener for reactive dyes.

The viscosity of the paste can be varied according to the application and the equipment. Thick pastes with short flow characteristics are useful when the extent of penetration into the fabric must be limited but thinner pastes with long flow are required for fine-patterned prints. For alginates containing small quantities of calcium, viscosity can be controlled by adding sequestering agents such as polyphosphates. However, these pastes are more likely to lose viscosity as shear rate increases and a paste that is less shear sensitive can be made using a high concentration of a lower-viscosity alginate. This latter kind of paste is especially useful for printing disperse dyes on synthetic fibers. Most alginate manufacturers can supply basic recipes for the different types of dyes and printing processes which are a useful starting point. The concentration can vary from 1.5% of high-viscosity alginate to 5% of low-viscosity alginate [9, 24, 25, 48, 51, 66, 69, 75, 79, 80].

3.3.7 Sizing Agent for Paper

Alginate is used as a surface sizing agent in the paper industry, where its addition to the normal starch sizing gives a smooth continuous film and a surface with less fluffing. The oil resistance of alginate film formed on the paper produces a better oil resistance so an improved gloss is obtained with high gloss inks. If papers or boards are to be waxed, alginate in the size can keep the wax mainly at the surface. The quantity of alginate used is usually 5–10% of the weight of starch in the size [74]. Zinc alginate can help improve the flameproof properties of paper products [73]. The flame resistance properties of alginate are particularly useful in coating cigarette paper, where the propagation of flame can be retarded and the remaining ash can be bound by the metal ions in alginate.

3.3.8 Coating for Welding Rods

During the production of welding rods, a coating is applied by using water-soluble sodium or potassium alginate, which acts as a flux to control the conditions in the immediate vicinity of the weld, such as temperature or oxygen and hydrogen availability. Alginate has the appropriate properties to meet the requirements for coating welding rods where the dry ingredients of the coating are mixed with sodium silicate (water glass) which gives some of the plasticity necessary for extrusion of the coating onto the rod and which also acts as the binder for the dried coating on the rod. Since the wet silicate has no binding action and does not provide sufficient lubrication to allow effective and smooth extrusion, the use of alginate can hold the damp mass together before extrusion and maintain the shape of the coating on the rod during drying and baking.

Water-soluble alginates such as sodium or potassium alginates are coated and are dried at moderate temperatures to form a protective film on the welding rod. With basic or low-hydrogen rods, calcium alginate, sometimes with a proportion of sodium alginate added, gives much better results. This is related to the high temperatures used to dry these rods (400–450 °C) which produces low moisture contents so that only very low hydrogen levels are found in the deposited weld metals. Water soluble alginates swell when wetted and as the water is driven out completely in this high-temperature drying, the alginates will contract and cracks will develop in the coating. When calcium alginate is mixed with sodium silicate, a small amount of sodium alginate forms around each particle of calcium alginate. This mixture is thixotropic, its viscosity is lowered when extrusion pressure is applied. It therefore acts as a good binder and extrusion lubricant. During the drying process, because the calcium alginate did not previously swell very much, it does not shrink appreciably and a more uniform coating result.

Figure 3.4 shows an illustration of alginate-coated welding rods. The quantities of alginates used are dependent on the type of welding rod being coated and the

Figure 3.4 An illustration of alginate-coated welding rods.

extrusion equipment being used. For water-soluble alginates, it may be 0.4–1.2% for low-hydrogen welding rods and 0.15–0.25% for acid and organic types. For the thixotropic alginates, manufacturers often find it more effective to use a mixture of calcium alginate and sodium alginate with a total alginate content of 0.4–0.6% for low hydrogen electrodes.

3.3.9 Binders for Fish Feed

The worldwide growth in aquaculture has led to the use of crude alginate as a binder in salmon and other fish feeds, especially moist feed made from fresh waste fish mixed with various dry components. Alginate binding can lower consumption by up to 40% and pollution of culture ponds is sharply reduced.

3.3.10 Biostimulants

Alginate is now increasingly used in the agricultural industry as a biostimulant. In the definition given by the European Biostimulants Industry Council, biostimulants contain substances and/or microorganisms whose function when applied to plants or the rhizosphere is to stimulate natural processes to enhance/benefit nutrient uptake, nutrient efficiency, tolerance to abiotic stress, and crop quality, independently of its nutrient content. Similarly, the USA 2018 Farm Bill defines biostimulant as a substance or microorganism that, when applied to seeds, plants, or the rhizosphere, stimulates natural processes to enhance or benefit nutrient uptake, nutrient efficiency, tolerance to abiotic stress, or crop quality and yield. Although alginate itself is not a nutrient, many studies have shown that it can promote plant growth through soil improvement, seed initiation, enhancement of healthy root growth, resisting biological and abiotic stress, and the protection of flowers and fruit. As a major component of seaweed fertilizers, alginate and its oligomers can effectively increase plant yield and improve product quality [8, 30, 41, 42, 64, 85, 87].

3.4 Summary

Alginate has many unique properties and has been used in many diversified industries. Traditionally, textile printing accounts for about half of the global market. Pharmaceutical and medical uses are now about 20% by value of the market and have stayed buoyant, driven by ongoing developments in controlled release technologies and the use of alginates in wound care applications. Food applications are worth about 20% of the market and are now growing due to their unique health benefits and gelling properties. In general, three principal grades of alginates are available with many variations of viscosity. The highest grades meet the requirements of the National Formulary (USA), the food grades meet the quality standards of the Food Chemicals Codex (USA), and the technical grades vary considerably in their color and water-insoluble solids such as cellulose.

References

1 Aldred, F.C. and Mosely, C.R. (1983). Man-made filaments and methods of making wound dressings containing them. US Patent 4,421,583.

2 Andres, C. (1987). Expanding applications for alginate technologies. *Food Process.* 48 (2): 30–32.

3 Anon, A. (1983). New concept for growing market: structured potatoes. *Food Eng.* 55 (5): 72–73.

4 Badwan, A. (1985). A sustained release drug delivery system using calcium alginate beads. *Drug Dev. Ind. Pharm.* 11: 239–256.

5 Bazett, M., Honeyman, L., Stefanov, A.N. et al. (2015). Cystic fibrosis mouse model-dependent intestinal structure and gut microbiome. *Mamm. Genome* 26 (5-6): 222–234.

6 Bowman, K.A., Aarstad, O.A., Nakamura, M. et al. (2016). Single molecule investigation of the onset and minimum size of the calcium-mediated junction zone in alginate. *Carbohydr. Polym.* 148: 52–60.

7 Burns, M.A., Kvesitadze, G.I., and Graves, D.J. (1985). Dried calcium alginate/magnetite spheres: a new support for chromatographic separations and enzyme immobilization. *Biotechnol. Bioeng.* 27 (2): 137–145.

8 Chandia, N.P., Matsuhiro, B., Mejias, E. et al. (2004). Alginic acids in *Lessonia vadosa*: partial hydrolysis and elicitor properties of the polymannuronic acid fraction. *J. Appl. Phycol.* 16: 127–133.

9 Christie, N.J. (1976). Thickeners for thermosol and space-dyeing. *Int. Dyer* 155: 19–23.

10 Connick, W.J. (1983). Controlled release of bioactive materials using alginate gel beads. US Patent 4,401,456.

11 Connick, W.J., Lee, R.E., and Rawson, J. (1984). Encapsulation with seaweed-based gels: a new process. *Agric. Res.* 32 (10): 8–9.

12 Cottrell, I.W. and Kovacs, P. (1980). Alginates. In: *Handbook of Water Soluble Gums and Resins* (ed. R.L. Davidson), 125–149. New York: McGraw-Hill.

13 Cox, J.P. (1982). Method for forming shaped products for human and/or animal consumption or as marine bait and products produced thereby. US Patent 4,362,748.

14 Dallyn, H., Falloon, W.C., and Bean, P.G. (1977). Method for the immobilization of bacterial spores in alginate gel. *Lab. Pract.* 26: 773–775.

15 De Lisle, R.C. (2007). Altered transit and bacterial overgrowth in the cystic fibrosis mouse small intestine. *Am. J. Physiol. Gastrointest. Liver Physiol.* 293 (1): G104–G111.

16 Draget, K.I. and Taylor, C. (2011). Chemical, physical and biological properties of alginates and their biomedical implications. *Food Hydrocoll.* 25: 251–256.

17 Fravel, D.R. (1985). Encapsulation of potential biocontrol agents in an alginate-clay matrix. *Phytopathology* 75: 774–777.

18 Frick, C. and Schuegerl, K. (1986). Continuous acetone-butanol production with free and immobilized *Clostridium acetobutylicum*. *Appl. Microbiol. Biotechnol.* 25: 186–193.

19 Garcia, M.A.S., Yang, N., and Quinton, P.M. (2009). Normal mouse intestinal mucus release requires cystic fibrosis transmembrane regulator-dependent bicarbonate secretion. *J. Clin. Invest.* 119 (9): 2613–2622.

20 Glicksman, M. (1969). *Gum Technology in the Food Industry*, 67–68. New York: Academic Press.

21 Glicksman, M. (1969). *Gum Technology in the Food Industry*, 239–273. New York: Academic Press.

22 Grizeau, D. and Navarro, J.M. (1986). Glycerol production by *Dunaliella tertiolecta* immobilized with calcium alginate beads. *Biotechnol. Lett.* 8: 261–264.

23 Gustafsson, J.K., Ermund, A., Ambort, D. et al. (2012). Bicarbonate and functional CFTR channel are required for proper mucin secretion and link cystic fibrosis with its mucus phenotype. *J. Exp. Med.* 209 (7): 1263–1272.

24 Hebeish, A. (1986). Technical feasibility of some thickeners in printing cotton with reactive dyes. *Am. Dyestuff Report.* 75 (2): 22–29.

25 Hilton, K.A. (1972). Sodium alginate in the textile industry. *Colourage* 19 (10): 65–68.

26 Johansen, A. and Flink, J.M. (1985). A novel method for immobilization of yeast cells in alginate gels of various shapes by internal liberation of calcium ions. *Biotechnol. Lett.* 7 (10): 765–768.

27 Johansen, A. and Flink, J.M. (1986). A new principle for immobilized yeast reactors based on internal gelation of alginate. *Biotechnol. Lett.* 8 (2): 121–126.

28 Johansen, A. and Flink, J.M. (1986). Immobilization of yeast cells by internal gelation of alginate. *Enzym. Microb. Technol.* 8 (3): 145–148.

29 Johansen, A. and Flink, J.M. (1986). Influence of alginate properties and gel reinforcement on fermentation characteristics of immobilized yeast cells. *Enzym. Microb. Technol.* 8 (12): 737–748.

30 Khan, W., Rayirath, U.P., Subramanian, S. et al. (2009). Seaweed extracts as biostimulants of plant growth and development. *J. Plant Growth Regul.* 28: 386–399.

31 King, A.H. (1983). Brown seaweed extracts (alginates). In: *Food Hydrocolloids* (ed. M. Glicksman), 115–188. Boca Raton, FL: CRC Press.

32 Klein, J. and Wagner, F. (1982). Preparation of immobilized enzymatically-active substance (on alginate gels). US Patent 4,334,027.

33 Klein, J., Stock, J., and Vorlop, K. (1983). Pore size and properties of spherical calcium-alginate biocatalysts. *Eur. J. Appl. Microbiol. Biotechnol.* 18: 86–91.

34 Kloosterman, J. and Lilly, M.D. (1986). Pilot-plant production of prednisolone using calcium alginate immobilized Arthrobacter simplex. *Biotechnol. Bioeng.* 28: 1390–1395.

35 Kobayashi, Y. (1986). Papers from seaweeds. Manufacture and applications of alginate fiber papers. *Tanpakushitsu Kakusan Koso* 31: 1066–1077.

36 Kobayashi Y, Matsuo R. *Manufacture of alginate paper*. Japanese Patent Kokai 174,499/86, 1986.

37 Kobayashi, Y., Matsuo, R., and Kawakatsu, H. (1986). Manufacture and physical properties of alginate fiber papers as an analysis model of cellulosic fiber papers. *J. Appl. Polym. Sci.* 31: 1735–1747.

38 Leigh, A.M. (1979). *Alginates in Food Production*, 11. London: Alginate Industries Ltd.

39 Lim, D.J. and Choi, C.Y. (1986). Citric acid production using immobilized yeast activated with calcium chloride-containing medium. *Korean J. Appl. Microbiol. Biotechnol.* 14: 285–292.

40 Littlecott, G.W. (1982). Food gels-the role of alginates. *Food Technol.* 34: 412–418.

41 Liu, H., Zhang, Y.H., Yin, H. et al. (2013). Alginate oligosaccharides enhanced *Triticum aestivum L.* tolerance to drought stress. *Plant Physiol. Biochem.* 62: 33–40.

42 Luan, L.Q., Nagasawa, N., Ha, V.T. et al. (2009). Enhancement of plant growth stimulation activity of irradiated alginate by fractionation. *Radiat. Phys. Chem.* 78: 796–799.

43 McDowell, R.H. (1960). Applications of alginates. *Rev. Pure Appl. Chem.* 10: 1–19.

44 McDowell, R.H. (1975). New developments in the chemistry of alginates and their use in food. *Chem. Ind.* 391–395.

45 McGhee, J.E., Carr, M.E., and St. Julian G. (1984). Continuous bioconversion of starch to ethanol by calcium alginate-immobilized enzymes and yeasts. *Cereal Chem.* 61: 446–449.

46 Means, W.J. and Schmidt, G.R. (1986). Algin/calcium gel as a raw and cooked binder in structured beef steaks. *Food Sci.* 51: 60–65.

47 Morimoto, K. (1985). Extrusion process for shrimp or crab meat analog products in a series of non-boiling gelling baths. US Patent 4,554,166.

48 Narkar, A.K. (1982). Thickeners for printing reactive dyes. *Indian Text. J.* 92 (7): 117–122.

49 Nordaård, C.T., Nonstad, U., Olderøy, M.O. et al. (2014). Alterations in mucus barrier function and matrix structure induced by guluronate oligomers. *Biomacromolecules* 15 (6): 2294–2300.

50 Nordgard, C.T. and Draget, K.I. (2011). Oligosaccharides as modulators of rheology in complex mucous systems. *Biomacromolecules* 12 (8): 3084–3090.

51 Obenski, B.J. (1984). Printing and print paste thickeners. *Am. Dyestuff Report.* 73: 15–16.

52 Onaka, T. (1985). Beer brewing with immobilized yeast. *Biotechnology* 3: 467–470.

53 Ooraikul, B. and Aboagye, N.Y. (1986). Synthetic food product. US Patent 4,582,710.

54 Padol, A.M., Draget, K.I., and Stokke, B.T. (2016). Effects of added oligoguluronate on mechanical properties of Ca-alginate-oligoguluronate hydrogels depend on chain length of the alginate. *Carbohydr. Polym.* 147: 234–242.

55 Pellico, M. (1983). Settable alginate compositions. US Patent 4,381,947.

56 Powell, L.C., Sowedan, A., Khan, S. et al. (2013). The effect of alginate oligosaccharides on the mechanical properties of gram-negative biofilms. *Biofouling* 29 (4): 413–421.

57 Prevost, H., Divies, C., and Rousseau, E. (1985). Continuous yoghurt production with *Lactobacillus bulgaricus* and *Streptococcus thermophilus* entrapped in calcium alginate. *Biotechnol. Lett.* 7: 247–252.

58 Pritchard, M.F., Powell, L.C., Menzies, G.E. et al. (2016). A new class of safe oligosaccharide polymer therapy to modify the mucus barrier of chronic respiratory disease. *Mol. Pharm.* 13 (3): 863–872.

59 Qin, Y. (2004). Gel swelling properties of alginate fibers. *J. Appl. Polym. Sci.* 91 (3): 1641–1645.

60 Qin, Y. (2005). The ion exchange properties of alginate fibers. *Text. Res. J.* 75 (2): 165–168.

61 Qin, Y. (2006). The characterization of alginate wound dressings with different fiber and textile structures. *J. Appl. Polym. Sci.* 100 (3): 2516–2520.

62 Qin, Y. (2008). Alginate fibres: an overview of the production processes and applications in wound management. *Polym. Int.* 57 (2): 171–180.

63 Qin, Y. (2008). The gel swelling properties of alginate fibers and their application in wound management. *Polym. Adv. Technol.* 19 (1): 6–14.

64 Qin, Y. (2022). *Marine Biostimulants*. Beijing: China Light Industry Press.

65 Qin, Y. and Gilding, D.K. (1996). Alginate fibers and wound dressings. *Med. Device Technol.* 11: 32–34.

66 Ramakrishnan, S. (1981). Textile thickening agents for reactive printing. *Colourage* 28: 9–11.

67 Rehg, T., Dorger, C., and Chau, P.C. (1986). Application of an atomiser producing small alginate gel beads for cell immobilization. *Biotechnol. Lett.* 8 (2): 111–114.

68 Rochefort, W.E., Rehg, T., and Chau, P.C. (1986). Trivalent cation stabilization of alginate gel for cell immobilization. *Biotechnol. Lett.* 8 (2): 115–120.

69 Rompp, W., Axon, G.L., and Thompson, T. (1983). Sodium alginate: a textile printing thickener. *Am. Dyestuff Report.* 72: 31–32.

70 Scheuble, M. and Munsch, P. (1983). Non-dusting and fast-wetting dental impression material. US Patent 4,394,172.

71 Schmidt, G.R. and Means, W.J. (1986). Process for preparing algin/calcium gel structured meat products. US Patent 4,603,054.

72 Schoutens, G.H. (1986). A comparative study of a fluidized bed reactor and a gas lift loop reactor for the IBE process. Part 3. Reactor performances and scale up. *J. Chem. Technol. Biotechnol.* 36: 565–576.

73 Sergeant, S.V. (1980). Improving the flame resistance of paper. *Converter* 17 (10): 14–15.

74 Sergeant, S.V. (1981). Applications of alginates in paper. *Converter* 18 (6): 28–29.

75 Shenai, V.A. and Saraf, N.M. (1981). Thickeners and additives in the printing paste. *Colourage* 28 (22): 44–48.

76 Shenouda, S.Y.K. (1983). Fabricated protein fiber bundles. US Patent 4,423,083.

77 Sletmoen, M., Maurstad, G., Nordgård, C.T. et al. (2012). Oligoguluronate induced competitive displacement of mucin-alginate interactions: relevance for mucolytic function. *Soft Matter* 8 (32): 8413–8421.

78 Tanaka, H., Matsumura, M., and Veliky, I.A. (1984). Diffusion characteristics of substrates in calcium alginate gel beads. *Biotechnol. Bioeng.* 26: 53–58.

79 Teli, M.D. and Chiplunkar, V. (1986). Role of thickeners in final performance of reactive prints. *Text. Dyer Print.* 19 (6): 13–19.

80 Teli, M.D., Shah, R., and Sinha, R. (1986). Developments in thickeners for reactive printing. *Text. Dyer Print.* 19 (7): 17–23.

81 Thomas, S. (2000). Alginate dressings in surgery and wound management – part 1. *J. Wound Care* 9 (2): 56–60.

82 Thomas, S. (2000). Alginate dressings in surgery and wound management – part 2. *J. Wound Care* 9 (3): 115–119.

83 Thomas, S. (2000). Alginate dressings in surgery and wound management – part 3. *J. Wound Care* 9 (4): 163–166.

84 Tramper, J. (1985). Immobilizing biocatalysis for use in synthesis. *Trends Biotechnol.* 3: 45–49.

85 Vera, J., Castro, J., Gonzalez, A. et al. (2011). Seaweed polysaccharides and derived oligosaccharides stimulate defense responses and protection against pathogens in plants. *Mar. Drugs* 9: 2514–2525.

86 Yang, N., Garcia, M.A., and Quinton, P.M. (2013). Normal mucus formation requires cAMP-dependent HCO_3^- secretion and Ca^{2+}-mediated mucin exocytosis. *J. Physiol.* 591 (18): 4581–4593.

87 Zhang, S., Tang, W., Jiang, L. et al. (2015). Elicitor activity of algino-oligosaccharide and its potential application in protection of rice plant (*Oryza sativa L.*) against *Magnaporthe grisea. Biotechnol. Biotechnol. Equip.* 29: 646–652.

4

The Production of Fibers From Alginate

4.1 Introduction

Since it was discovered by Stanford [56] in 1881, alginate has been used in a wide variety of industries, such as food, textile printing, paper, pharmaceuticals, and many others [39]. As a water-soluble polymer, alginate is an excellent gel-forming material capable of holding a large amount of water, and in recent years, alginate has been widely used in the wound management industry as a novel material for the manufacture of "moist healing" products such as gels, foams, and fibrous nonwoven dressings used to cover wounds. In these applications, the alginate-based materials are used either in the dry form to absorb wound fluid or to donate water to a dry wound when they are used in the hydrated gel form. In both cases, the interaction of the alginate materials and wound surface creates a moist environment. It has been shown that when wounds are kept in a moist but not wet condition, the migration of epithelial cells from the edge of the wound to the wounded area is faster than when the wounds are kept in a dry state. Modern "moist healing" wound dressings aim to create the moist condition that can facilitate optimum healing [15].

Alginate fibers are particularly useful as raw materials for the production of highly absorbent wound dressings. In this particular field, the properties of alginate fibers are unparalleled in many respects. First, as a natural polymer, alginate is nontoxic and safe to use on wound surfaces and in cavities. Second, when the water-insoluble calcium alginate fiber is placed in contact with wound exudates, the calcium ions in the fibers exchange with sodium ions in the body fluid, with the calcium ions released from the fibers acting as a hemostatic agent. Third, as calcium alginate fibers slowly turn into sodium alginate fibers, they absorb a large quantity of exudates and turn into a fibrous gel, which helps to keep a moist interface on the wound surface. Forth, as a natural polymer, alginate is a renewable resource with unlimited supply in nature [41, 51]. In the fiber form, it is also possible to process the alginate fibers into woven, nonwoven, knitted, and various types of composites to address specific wound management problems.

Alginate Fibers and Wound Dressings: Seaweed Derived Natural Therapy, First Edition. Yimin Qin.

4.2 The Properties of Alginate as a Fiber-Forming Polymer

Alginate can be extracted from brown seaweeds by a treatment with aqueous alkali solutions, typically NaOH and Na_2CO_3, in which the natural alginate in various salt forms is converted into water-soluble sodium alginate [36]. After filtration, the sodium alginate in solution can be precipitated by the addition of calcium chloride. After further purification and conversion, water-soluble sodium alginate power is produced, which can be used as the raw material for the production of alginate fibers. It is worth mentioning here that alginate fibers account for only a small percentage of alginate extracted from brown seaweeds [7], with the majority of the global output used in food ingredients, textile printing, pharmaceuticals, and other industries.

The British Isles are rich in seaweed resources and after the discovery of alginate by Stanford in 1881, several companies were established to commercialize the production of alginate from brown seaweeds, such as British Algin Company Ltd. (1885), Blandola Ltd. (1908), and Liverpool Borax Ltd. (1909). The first large-scale commercial production of alginate was by Kelco based in San Diego, which was established by F C Thornley in 1929. There were many attempts to make industrial products from alginate, for example glass paper by Cefoil Ltd. which was established in 1934. During the Second World War, the British government initiated a project to produce alginate fibers for use as military camouflage textiles, since the import of linen from India was disrupted by the war [64]. Although calcium alginate fibers were successfully developed, these fibers were later shown to be not suitable for the production of conventional textile products, since they are generally not stable in acidic and alkaline conditions, and more importantly, the rapid development of the chemical fiber industry resulted in a large variety of new fibers that are cheaper and better than alginate fibers. However, the discovery of the "moist healing" mechanism and the commercial success of "moist healing" wound management products in the 1980s stimulated the widespread use of alginate wound dressings, which revived the worldwide interest in alginate fibers until the present time.

As a raw material for fiber production, alginate is a linear polymeric acid composed of 1, 4-linked β-D-mannuronic acid (M) and α-L-guluronic acid (G) residues. These two acid residues are stereochemically very different as a result of their difference at the C-5 position. Alginates extracted from different species of seaweeds differ in their M and G contents, resulting in differences in the physical properties of the respective alginate products. Table 4.1 shows the M and G contents of alginate extracted from seaweeds commonly used for the commercial production of alginate [22, 23, 27].

Because its polymeric structure contains two types of monomer acids, alginate can be regarded as a block copolymer of β-D-mannuronic acid and α-L-guluronic acid. It has been shown that the polymer chain is made up of three kinds of blocks, where the GG blocks contain only units derived from L-guluronic acid, the MM blocks are based entirely on D-mannuronic acid, and the MG blocks consist of alternating units from D-mannuronic acid and L-guluronic acid [25, 40]. It was demonstrated from X-ray studies on the fibers of polymannuronic acid and those from alginic

Table 4.1 Percentages of mannuronic acid and guluronic acid, and M/G ratios of alginate extracted from brown seaweeds.

Type of seaweeds	Mannuronic acid content (%)	Guluronic acid content (%)	M/G ratio
Ascophyllum nodosum	60.0	40.0	1.5
Laminaria digitata	59.0	41.0	1.43
Saccharina japonica	69.3	30.7	2.26
Macrocystis pyrifera	61.0	39.0	1.56
Laminaria hyperborea, fronds	56.0	44.0	1.28
Laminaria hyperborea, stems	30.0	70.0	0.43

acid rich in guluronic acid content that the spacings along the fiber axis are 10.35 Å for mannuronic polymer and 8.72 Å for guluronic polymer [5, 6]. Conformations of the uronic acid units in agreement with these spacings are Cl for the 1,4-linked β-D-mannuronic acid units and lC for the 1,4-linked α-L-guluronic acid units. The MM block is linked diequatorially at C-1 and C-4, it is a relatively straight polymer, like a flat ribbon, while the GG block is formed from diaxial groups at both C-1 and C-4, so the resulting chain is buckled. Figure 4.1 shows the schematic representation of the stereochemical structures of MM and GG blocks.

In addition to the G and M contents, it has been shown that the physical properties of alginates also depend on the relative proportions of the three types of blocks [24, 53, 54]. For example, the formation of gels by the addition of calcium ions mainly involves the GG blocks. Hence, under otherwise same conditions, the higher the

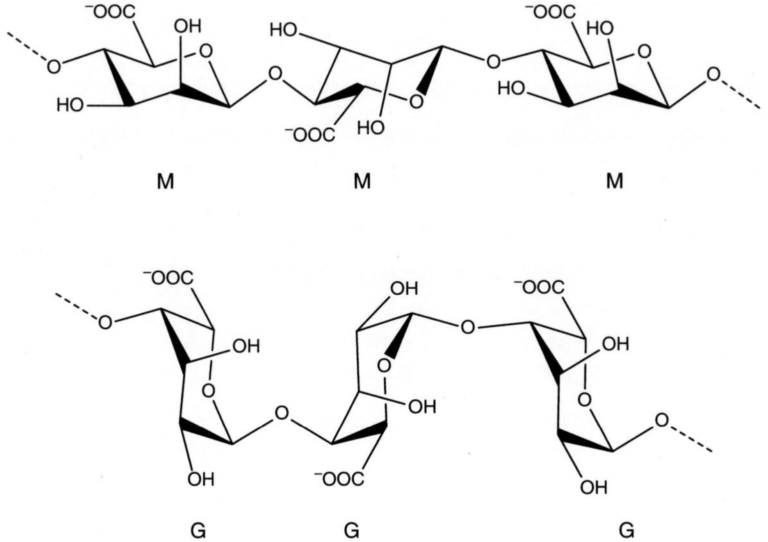

Figure 4.1 Stereochemical structures of MM and GG blocks of alginic acid.

Table 4.2 Percentages of the three principal types of block structures in alginate extracted from several brown seaweeds.

Type of seaweeds	MM segments (%)	GG segments (%)	MG/GM segments (%)
Ascophyllum nodosum	38.4	20.7	41.0
Laminaria digitata	49.0	25.0	26.0
Saccharina japonica	36.0	14.0	50.0
Macrocystis pyrifera	40.6	17.7	41.7
Laminaria hyperborea, fronds	43.0	31.0	26.0
Laminaria hyperborea, stems	15.0	60.0	25.0

GG contents, the greater the gel strength. Table 4.2 shows the percentages of the three principal types of block structures in alginate extracted from several brown seaweeds.

During the production of alginate fibers, the production process and the product performances are closely related to the properties of alginate, and it is important to quantify the relative proportions of the uronic acids. Various methods have been developed to measure the ratio of mannuronic acid to guluronic acid (the M/G ratio), as well as the MM, GG, and MG/GM contents of alginate samples [4]. In particular, the accurate determination of G, M, MM, GG, and MG/GM contents can be measured by ^{1}H-n.m.r and ^{13}C-n.m.r [21].

4.3 Preparation of the Spinning Solutions

In the wet-spinning process in which alginate is converted from powder into fiber form, alginate is first dissolved in water to form a homogeneous solution. Sodium alginate is typically used as the raw material, since it is easily soluble in water. In preparing a spinning solution, it should be pointed out that the properties of the alginate fibers depend on a number of factors, such as the molecular structure and molecular weight of the alginate, the coagulation bath composition, and the processing temperatures and speeds. As the first step in the production process, the spinning solution has an important effect on both the production efficiency and the product performance. Some of the factors involved in the preparation of an alginate spinning solution are discussed below.

4.3.1 Molecular Weight of the Alginate Powder

In theory, a higher molecular weight would result in a higher level of inter-chain bonding and a higher fiber strength. However, during the preparation for spinning solution, it should be noted that as the molecular weight increases, solution viscosity increases sharply, making it difficult to dissolve a high concentration of alginate in a given amount of water. Regarding the molecular weight of alginate, although it is a natural polymer, manufacturers can control the molecular weight (or degree of polymerization, DP) by varying the severity of the extraction conditions to produce products with viscosities in a 1% solution ranging from 10 to 1000 mPa s, with a DP range of 100–1000 units. In order to extrude the alginate solutions through spinneret holes to form filaments, the spinning solution typically has a viscosity in the range of 10 000–20 000 mPa s, which can be obtained either from a high concentration solution made of alginate with a low molecular weight or a low concentration with a high molecular weight. For the production of alginate fibers, in order to balance the production efficiency with product performance, the raw material typically has a viscosity in 1% solution in the range of 40–100 mPa s.

4.3.2 Concentration of the Spinning Solution

The viscosity of the alginate solution increases sharply as the concentration increases. In order to achieve high productivity, high concentrations are normally preferred in the wet-spinning process for the high production efficiency and the generally good fiber properties, since as-made fibers from a spinning solution with a high solid content have a higher wet strength and are relatively easy to carry out the stretching and washing processes. However, high solution viscosities make it difficult to remove bubbles in the solution brought in during the mixing process. For practical reason, wet-spinning processes typically use aqueous solutions containing 5–6% sodium alginate as the spinning solution.

4.3.3 Temperature of the Spinning Solution

The viscosity of alginate solutions decreases as temperature increases which improves processing. However, if alginate solutions are kept above 50 °C for several hours, depolymerization may occur, giving a permanent loss of viscosity and molecular weight. During the preparation of alginate spinning solutions, it is usual to use high-shear mixers and the solution temperature is raised by the heat generated from high levels of shearing. This is beneficial since the reduced viscosity helps the bubbles to rise from the spinning solution. However, a prolonged storage of the alginate solution at high temperatures will have a detrimental effect on its molecular weight. It is common to store the spinning solution at room temperature prior to extrusion.

4.3.4 pH of the Spinning Solution

Aqueous alginate solutions are stable at room temperature over the range of pH 5–11. For the wet-spinning of alginate, the spinning solution is usually prepared by dissolving sodium alginate in de-ionized water, with pH of around 7.

4.4 The Production of Calcium Alginate Fibers

Figure 4.2 shows a schematic illustration of a small wet spinning production line where the production of calcium alginate fibers undergoes a series of steps from

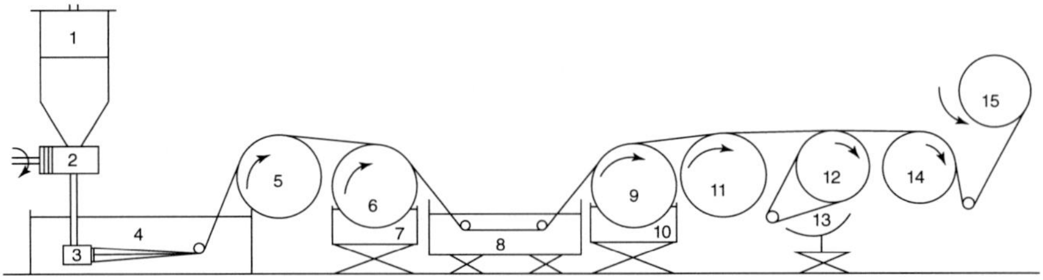

Figure 4.2 Schematic illustration of a small wet spinning production line, (1) spinning solution; (2) metering pump; (3) spinneret; (4) coagulation bath; (5) take-up roller I; (6) take-up roller II; (7) water bath; (8) hot water stretching bath; (9) stretching roller I; (10) acetone washing bath; (11) stretching roller II; (12) advancing roller I; (13) heater; (14) advancing roller II; (15) winding up unit. Adapted with permission, Qin [45].

the extrusion of spinning solution to winding up of the resultant fibers. In terms of the manufacturing process, calcium alginate fibers are made in one of the most basic spinning processes in which the spinning solution is made by dissolving sodium alginate powder in water and after degassing to remove the bubbles in the solution, the sodium alginate solution is extruded through fine spinneret holes into a calcium chloride bath, whereby sodium alginate is precipitated out in filament form as calcium alginate fiber. The as-made fibers are then stretched, washed, and dried to produce calcium alginate fibers.

It should be pointed out that solutions of sodium alginate can react with many di- and tri-valent cations to form gels, hence it is possible to use a variety of metal ions to precipitate sodium alginate solution during the wet-spinning process. In the production of alginate fibers, calcium ion has found the greatest popularity as the divalent ion for gel formation, mainly because its salts are cheap, readily available, and nontoxic. Zinc chloride has also been used for the production of zinc alginate fibers.

Figure 4.3 shows an illustration of the binding of calcium ion by a GG block in alginate, resulting in the so-called egg-box structure [20]. During the extrusion process where sodium alginate is extruded into a calcium chloride bath, the buckled chain of guluronic acid units acts as a two-dimensional analog of a corrugated egg-box with interstices in which the calcium ions may pack and be coordinated, and while calcium ions help to hold the alginate molecules together, their polymeric nature and their aggregation bind the calcium ion more firmly, resulting in a firm gel structure. During gel formation, the structure of the guluronic acid block gives distances between carboxyl and hydroxyl groups which allow a high degree of coordination of the calcium.

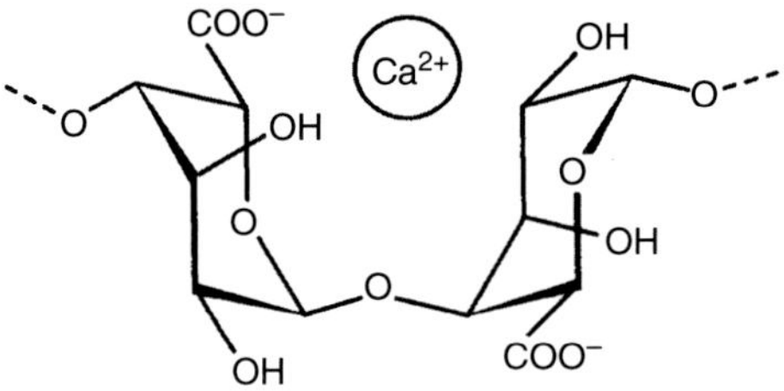

Figure 4.3 Binding of calcium ion by a GG block in alginate.

Figure 4.4 An illustration of the extrusion of sodium alginate spinning solution into a calcium chloride coagulation bath.

Figure 4.4 shows an illustration of the extrusion of sodium alginate spinning solution into a calcium chloride coagulation bath. In this wet-spinning process for calcium alginate fibers, calcium ions from the coagulation bath diffuses into the fiber to form a swollen fibrous gel. Thomas et al. [61] studied the diffusion of calcium ions during wet-spinning process. Their results showed that the gelling time of calcium alginate fiber varied linearly with alginate concentration, increased markedly with fiber radius, and decreased with increase in calcium concentration. It is interesting that the gelling time is independent of the guluronic and mannuronic acid contents. Gelling time for a given sodium alginate filament can be calculated in the following empirical equation:

$$T = ER^2/4DC + ER/2KC$$

where

T: Gelling time (s)
R: Radius of fiber (m)
E: Stoichiometric constant (kmol/m^3)
D: Diffusivity (m^2/s)
C: Calcium chloride solution strength (kmol/m^3)
K: Mass transfer coefficient (m/s^1)

Experimental results showed that when the fiber radius increased from 0.65 to 3.0 mm, the time required for gelling increased from 200 to 3500 seconds. For a wet-spun filament, the typical radius is about 100 μm, and the time required for coagulation is about five seconds.

Speakman and Chamberlain [55] made a detailed study of the production of calcium alginate fibers. They made fibers from a series of alginate with different molecular weight, characterized with different time of fall of the steel ball. As can be seen in Table 4.3, when the ball falling time increased from 2.0 to 174.0 seconds, indicating a big change in molecular weight, the fiber strength varied between 1.45 and 1.68 g/d. This shows that the properties of the alginate fibers were not seriously affected by the changes in the molecular weight of alginate.

Table 4.4 shows the properties of calcium alginate fibers made from spinning solutions with different solid contents. The best result was obtained when the solid content was 3.92%. As the concentration increased, solution viscosity became too high and it was difficult to spin the fiber. However, fibers made with a high concentration of alginate had a good luster, with the cross-section assuming a round shape. As can be seen in Figure 4.5, when the solid contents increased from 2.25% to 8.88%, the cross-section of the fibers changed from an irregular shape to a round shape.

Table 4.3 Properties of calcium alginate fibers made from sodium alginate with different molecular weights.

Ball falling time at 25 °C (seconds)	Solid content of the spinning solution (%)	Fiber extension at break (%)	Fiber tenacity (g/d)
2.0	4.07	9.2	1.48
17.6	3.92	11.1	1.51
20.9	4.07	12.9	1.45
42.1	3.71	12.6	1.68
57.7	3.80	12.5	1.65
174.0	3.99	10.5	1.60

Table 4.4 Properties of calcium alginate fibers made from spinning solutions with different solid contents.

Solid contents (%)	Ball falling time at 25 °C (seconds)	Fiber extension at break (%)	Fiber tenacity (g/d)
2.25	1.83	7.9	1.44
3.92	17.6	11.1	1.51
5.93	56.5	13.1	1.37
7.48	159.3	12.5	1.23
8.88	610.0	14.5	1.23

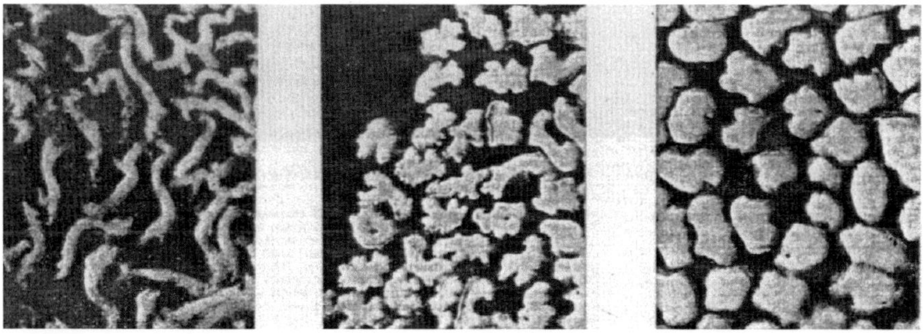

Figure 4.5 Cross-sectional shapes of alginate fiber made from alginate solutions with different solid contents, from left to right, 2.25%, 5.93%, and 8.88%. Adapted with permission, Qin [45].

4.5 The Production of Calcium Sodium Alginate Fibers

Alginate fibers are mainly used for the production of wound dressings, where absorption of wound fluid is a key performance criterion. During the development of alginate wound dressings, calcium alginate fibers were chemically treated to convert them into a mixed salt containing both calcium and sodium ions, where the calcium ions give the fiber wet integrity, while the sodium ions enhance the fiber absorbency. In the process of making calcium sodium alginate fibers, the calcium alginate fibers are first washed with hydrochloric acid to replace part of the calcium ions with hydrogen ions. The hydrogen ions are then replaced with sodium ions by a treatment with Na_2CO_3 or NaOH. The resultant fiber contains both calcium and sodium alginate. Because sodium alginate is water soluble, the fibers become more and more absorbent when more and more sodium ions are introduced into the fibers [52].

In order to make alginate fibers with different amounts of sodium ions, the calcium alginate fibers can be treated with aqueous solutions containing different amount of Na_2SO_4. Na_2SO_4 is used because the solubility of $CaSO_4$ in water is only 0.209 g per 100 ml at 30 °C, hence it can easily replace calcium ions from the alginate fibers. Table 4.5 shows the calcium and sodium contents after the calcium alginate

Table 4.5 Effect of Na_2SO_4 concentration on fiber calcium content and gel swelling ratio.

Na_2SO_4 concentration	Fiber Ca (II) content (%)	Fiber Na (I) content (%)	% Alginic acid as Na (I) salt (%)	Gel swelling ratio in water
Control	8.35	0.20	2.0	1.8 ± 0.15
0.1%	6.55	1.25	14.2	6.5 ± 0.45
0.2%	6.45	1.6	17.7	8.5 ± 0.65
0.5%	5.65	1.85	22.2	11.1 ± 0.80
0.7%	5.95	2.65	27.9	21.0 ± 1.50

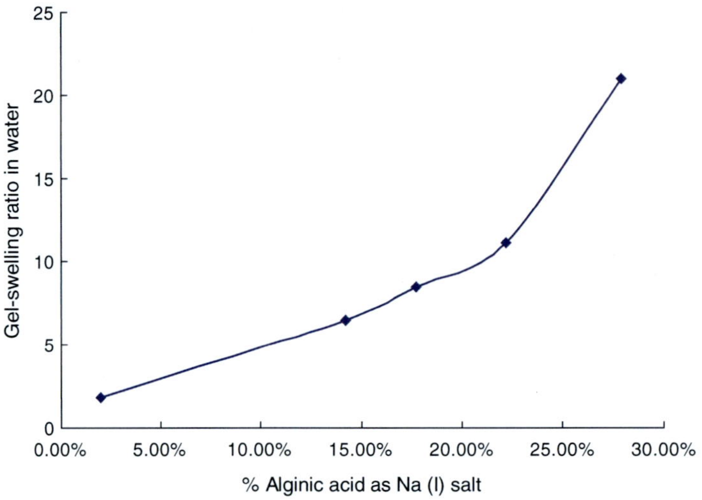

Figure 4.6 Effect of sodium content on the gel swelling ratio of alginate fibers.

fibers were treated with aqueous solutions containing various levels of Na_2SO_4. It is clear that as the Na_2SO_4 concentration increases, more and more calcium ions were replaced by sodium ions. Gel swelling results show that whilst the original sample swelled only a little in water, with a gel swelling ratio of 1.8, there were significant increases in the gel swelling ratio as more and more calcium alginate was converted into sodium alginate. As can be seen in Figure 4.6, gel swelling ratio rose as the fiber sodium content increased, with the sample treated with 0.7% Na_2SO_4 showing a gel swelling ratio of 21.

Figure 4.7 shows the photomicrographs of alginate fibers with various levels of sodium contents. It is clear that the sodium contents in the alginate fibers have a significant effect on the absorption and gelling properties of the fiber [42].

4.6 The Production of Sodium Alginate Fibers

Sodium alginate fibers can be produced by extruding the sodium alginate solution into an organic solvent bath. Kobayashi et al. [32] made extensive investigations and found that it is possible to produce sodium alginate fibers by extruding an aqueous sodium alginate solution into a coagulation bath containing a large quantity of a hydrophilic organic solvent in which sodium alginate is insoluble. The prompt displacement of water in the spinning solution by the organic solvent produces continuous filaments of sodium alginate.

To prepare sodium alginate fibers, a 5% aqueous sodium alginate solution is first made by adding the sodium alginate powder to distilled water and stirring for about four hours. The solution is filtered through a 200-mesh filter cloth to prepare a dope. After being stored overnight under reduced pressure, the dope was extruded at a rate of 12 g per minute into a coagulating bath of varied non-solvents at 18 °C, through

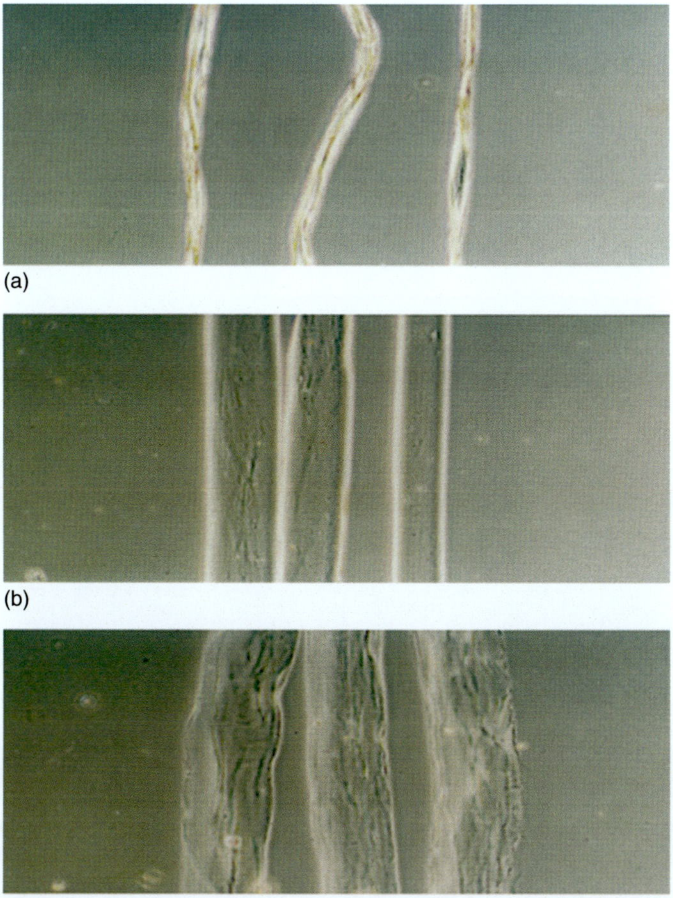

(a)

(b)

(c)

Figure 4.7 Photomicrographs of alginate fibers with (a): 2.0%; (b): 17.7%; (c): 27.9% sodium alginate, wet in water (× 200).

a spinneret plate with 1000 holes, each having a diameter of 0.1 mm. The resultant fibers were rolled at a rate of 0.5 m per minute onto a godet roller. The fibers were immersed in acetone, and then dried at 100 °C.

Table 4.6 shows the properties of the resultant sodium alginate fibers made with different types of coagulation baths. It appears that although sodium alginate fibers can be made by extruding their solution into various organic solvents, the fiber tenacity is generally low. The highest tenacity of 0.338 g/d obtained with isopropanol is much lower than a typical fiber tenacity of about 1.8 g/d for calcium alginate fiber.

In another experiment, sodium alginate solution was extruded into acetone at 13 °C at a rate of 16.4 g per minute through a spinneret plate with 1000 holes, each having a diameter of 0.1 mm. The resultant fibers were rolled onto a godet roller with a diameter of 11.2 cm at varied rotation rates. Table 4.7 shows the fineness and

Table 4.6 Properties of sodium alginate fibers made with different types of coagulation bath.

Coagulation bath	Fiber fineness (denier)	Fiber tenacity (g/d)
Methanol	5.463	0.236
Ethanol	4.734	0.250
Isopropanol	4.491	0.338
Acetone	4.410	0.312

Table 4.7 Properties of sodium alginate fibers made with different take-up speeds.

Rotation rate (rpm)	Fiber fineness (denier)	Fiber tenacity (g/d)
4	3.90	0.447
8	2.94	0.811
16	1.54	0.802
24	0.95	0.969
32	0.71	1.100

strength of the sodium alginate fibers made with different take-up speeds. It is clear that as the take-up speed increases, the fibers were drawn finer and since there is more orientation of the polymeric molecules along the fiber axis, there was also a significant improvement in the fiber strength [45].

4.7 The Production of Alginic Acid Fibers

Alginic acid fibers can be made by either a direct extrusion of sodium alginate solution into an aqueous bath containing sulfuric acid, where sodium alginate precipitates out as water-insoluble alginic acid, or by the conversion of calcium alginate fiber into alginic acid fiber in fiber form. Because alginic acid is insoluble in water, the removal of calcium ions with hydrochloric acid can be carried out in an aqueous media. The reaction proceeds according to the following equation:

$$\text{Calcium alginate} + 2HCl \rightarrow \text{Alginic acid} + CaCl_2$$

In this process, the alginate fibers remain in the solid state, while the calcium ions are washed into the solution. It has been shown that the calcium ions in the calcium alginate fibers can be easily washed off from the fibers by treating the fibers with an excess amount of 0.5 mol/L aqueous HCl solution at room temperature. No calcium ions were found in the fibers after 20 minutes of treatment [52].

4.8 The Production of Zinc Alginate Fibers

In clinical practice, zinc ions can generate immunomodulatory and antimicrobial effects, as well as activation of matrix metalloproteinases that facilitate auto debridement and keratinocyte migration [2], and topical applications of zinc-containing wound dressings are common in wound management. Topical zinc oxide has been found to aid treatment of leg ulcers and pressure sores [1, 3].

Since alginate is a polymeric acid, it is easy for alginate fibers to carry zinc ions by forming zinc alginate. Zinc alginate fibers can be made by a direct extrusion of sodium alginate solution into a zinc chloride coagulation bath, or by treating calcium alginate fibers with aqueous zinc chloride solution, where after the ion-exchange between calcium ions in the fiber and zinc ions in the solution, the fiber becomes a mixed calcium and zinc alginate.

Table 4.8 shows the zinc content and zinc release after calcium alginate fibers were treated with zinc chloride. It is clear that after the calcium alginate fibers were treated with $ZnCl_2$ solutions, both the high G and high M calcium alginate fibers contain a significant amount of zinc ions. When placed in solution A (a solution with similar levels of calcium and sodium ions to body fluid, defined in the British Pharmacopoeia as an aqueous solution containing 142 mmol of sodium chloride and 2.5 mmol of calcium chloride), the zinc ions can be replaced by sodium ions in the contacting solution in a similar way to calcium ions, resulting in the release of zinc ions. Since high G alginate binds more firmly with zinc ions, the release of zinc from zinc-containing alginate fibers is easier with high M-type alginate fiber. After 48 hours in contact with solution A, the zinc contents were 640 and 449 ppm, respectively, in the high M and high G alginate samples [42].

4.9 The Production of Alginate Fibers Containing Pectin and Carboxymethyl Cellulose

By blending other water-soluble materials into the spinning solution, it is possible to improve the absorption capacity of the alginate fibers by breaking the regular

Table 4.8 Zinc content and zinc release after calcium alginate fibers were treated with zinc chloride.

Test criteria	High G alginate fiber	High M alginate fiber
Zn content after $ZnCl_2$ treatment	7.75%	10.5%
Ca content after $ZnCl_2$ treatment	2.75%	3.05%
Zn content in contact solution, after 1 hour (ppm)	461	495
Zn content in contact solution, after 24 hours (ppm)	441	632
Zn content in contact solution, after 48 hours (ppm)	449	640

structure of the fibers and making it easy for the ion exchange to take place. Qin and Gilding [49] prepared fibers from a mixture of alginate, CMC, and pectin. A spinning dope was prepared by mixing 12 kg of sodium alginate, 1.5 kg of CMC, and 1.5 kg of high methoxy pectin in 235 l of water. After storage at room temperature for two days to remove the bubbles, fibers were produced by extruding the dope through a 40 000-hole spinneret at 12 m/min. The as-spun fibers were taken up at 7.2 m/min and then stretched at 80 °C to 9 m/min. The fibers were then washed with water before they were dried by first passing the fibers through an acetone bath and then drying them with heated air. Finally, the dry tow was crimped and cut to produce staple fibers. In another example, fibers were made from alginate and CMC by first mixing 13.5 kg of sodium alginate and 1.5 kg of CMC in 235 l of water. The fibers could be spun and carded into a nonwoven dressing.

Figure 4.8 shows an illustration of the swelling process of alginate/CMC fiber when in contact with sodium ions. As anionic water-soluble polymers, sodium alginate and sodium carboxymethyl cellulose can mix evenly in the spinning solution. When made into fibers, the sodium carboxymethyl cellulose molecules are dispersed in the alginate matrix. By disrupting the regular structure, it makes it easy for ion exchange to take place and the alginate/CMC polyblend fibers and wound dressings are much more absorbent than the pure alginate fibers and dressings. As shown in Table 4.9, under the same test conditions, the alginate/CMC composite dressing has an absorbency of 20.35 g/g, while the pure alginate dressing has an absorbency of 14.27 g/g.

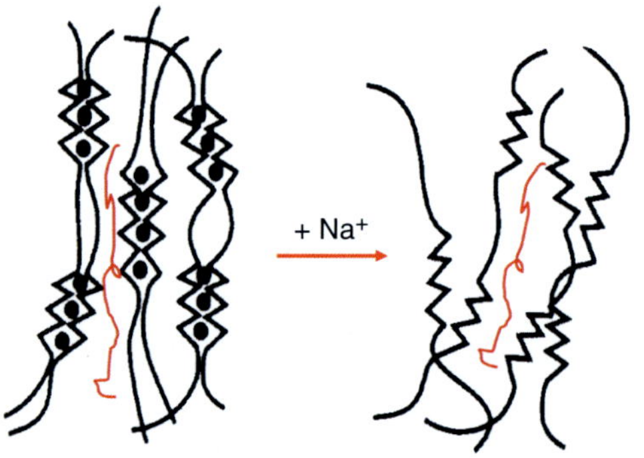

Figure 4.8 An illustration of the swelling process of alginate/CMC fiber when in contact with sodium ions.

Table 4.9 A comparison of the absorption properties of alginate/CMC composite dressing and pure alginate dressing.

Tests	Alginate/CMC composite dressing	Pure alginate dressing
Absorbency (g/g)	20.35 ± 0.75	14.27 ± 0.41
Gel swelling ratio in water (g/g)	4.25 ± 0.25	1.85 ± 0.13
Gel swelling ratio in saline (g/g)	9.65 ± 1.42	5.23 ± 0.42

4.10 The Production of Silver-Containing Alginate Fibers

Silver has a long history as an antimicrobial agent, especially in the treatment of burns [29, 30]. While metallic silver is relatively inactive, silver ions are effective against a wide range of bacteria. When low concentrations of silver ions accumulate inside cells, they can bind to negatively charged components in proteins and nucleic acids, thereby effecting structural changes in bacterial cell walls, membranes, and nucleic acids that affect viability. Although silver is highly effective against bacteria and other microorganisms, it has a limited toxicity to mammalian cells, making it an excellent antimicrobial agent for clinical uses [33].

In recent years, silver has been gaining importance in the wound management industry, and a number of silver-containing wound dressings have been developed. These function by the sustained release of low concentrations of silver ions over time, and generally appear to stimulate healing, as well as inhibiting microorganisms. A number of laboratory studies have shown the excellent antimicrobial performances of the silver-containing wound dressings [19, 34, 59, 60].

Since alginate wound dressings are highly absorbent, they are mainly used on highly exuding wounds where microbial infection is common. By incorporating silver ions into alginate fibers, it is possible to obtain highly absorbent wound dressings with good antimicrobial properties.

As a polymeric acid, alginate can form salt with silver ions. However, unlike calcium alginate, which is highly insoluble in water, silver is a monovalent ion and when sodium alginate solution is extruded into a silver nitrate solution, it is difficult to form silver alginate fiber. A mixed solution of calcium chloride and silver nitrate can be used to produce fibers that are a mixture of calcium alginate and silver alginate. Table 4.10 shows the calcium and silver contents of alginate fibers made with a mixed coagulation bath containing calcium and silver ions [35].

In order to attach silver ions in the alginate fibers, calcium alginate fibers can be treated with aqueous solutions of silver nitrate. The silver ions in the solution

Table 4.10 Preparation conditions and properties of the calcium silver alginate fiber.

Test criteria	Sample no.1	Sample no.2
Alginate content of the spinning solution	6%	6%
Coagulation duration	30 seconds	600 seconds
Fiber silver content	5.12%	7.30%
Fiber calcium content	4.98%	6.18%
Ratio between silver/calcium	1.03	1.18
Fiber strength (cN/dtex)	1.09	1.15
Extension at break (%)	10.5	8.9

exchange with calcium ions in the fiber, resulting in the formation of calcium alginate fiber containing silver ions. These fibers are highly antimicrobial. However, due to the oxidative power of the silver ions, they are sensitive to light exposure and can become dark to black in appearance.

Adding particles of water-insoluble silver compounds into the alginate fiber is one way to avoid oxidation and maintain the white physical appearance that is highly desirable for a biomedical material. Le et al. [35] developed a method to incorporate silver sulfadiazine (SSD) into alginate fibers by mixing the water-soluble sodium sulfadiazine with sodium alginate to form a spinning solution, which was then extruded into a 2% calcium chloride solution containing silver nitrate. During the fiber-forming process, sodium alginate reacts with calcium ions to form the filament, whilst sodium sulfadiazine reacts with silver ions to form SSD, which is deposited inside the fiber structure. Alternatively, after sodium sulfadiazine and sodium alginate are dissolved to form a spinning solution, silver nitrate is added into this solution before extrusion. In this process, the SSD particles formed through the reaction between sodium sulfadiazine and silver nitrate are dispersed in the spinning solution. When extruded to form fiber, the SSD particles are embedded in the fibers.

As been mentioned before, although silver is a highly effective broad-spectrum antimicrobial agent, it is also highly oxidative to organic materials. Skin discoloration and irritation associated with the use of silver nitrate is well known. In order to protect the host material from oxidation and discoloration, some novel silver-containing compounds have been developed in recent years and these have been made into fine particles that can be blended with fiber-forming polymers during extrusion.

AlphaSan RC5000 is a silver sodium hydrogen zirconium phosphate. This microbiologically active ingredient is a synthetic inorganic polymer. Under scanning electron microscope, it resembles cube-shaped crystals, with an average particle size of about one micron. It consists of a three-dimensional repeating framework of sodium hydrogen zirconium phosphate, with many equally spaced cavities containing silver. Silver (at 3.8% by weight) provides the main antimicrobial properties, while the framework matrix acts to distribute silver evenly without clumping or pooling throughout the individual fibers where the AlphaSan particles are added.

When AlphaSan RC5000 is mixed with sodium alginate solution, the fine particles can be evenly distributed in the spinning solution under a high rate of shearing. Because the particles are very fine, they can be suspended uniformly while the solution is extruded to form fibers. Since the sodium hydrogen zirconium phosphate framework prevents the silver ions from oxidizing the alginate, this type of silver-containing alginate fiber remains white even after sterilization through irradiation [50].

When alginate fibers containing AlphaSan RC5000 particles are in contact with wound exudates, the silver ions can be released into the wound exudate by three mechanisms. First, there is an ion exchange between the silver ions in the fiber and the sodium and calcium ions in the wound fluid. Second, silver ions can be chelated by protein molecules in the wound fluid. Third, AlphaSan particles attached on the

Table 4.11 Silver concentrations in contact solutions.

Duration of contact	Silver concentration in contact solution (ppm)	
	Normal saline	Human serum
30 minutes	0.50	2.18
48 hours	0.40	2.74
7 days	1.32	3.74

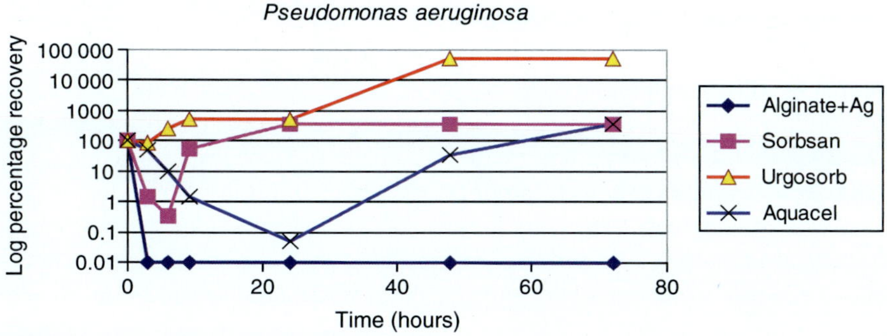

Figure 4.9 The antimicrobial action of silver-containing alginate fibers against *Pseudomonas aeruginosa*. Adapted with permission from Qin [43].

surface of the fibers can also be detached from the fibers and get into the wound exudate.

Table 4.11 shows the silver ion concentration when alginate fiber containing 1% AlphaSan RC5000 is placed in contact with normal saline or human serum. It can be seen that the silver ions are slowly released into the solution, acting as an antimicrobial agent. More silver ions can be seen released into human serum, suggesting the high silver binding abilities of the protein components in the wound exudates [43].

Figure 4.9 shows the antimicrobial action of silver-containing alginate fibers against *Pseudomonas aeruginosa*. There was 100% reduction in bacteria count within five hours when the fibers were placed in contact with solutions containing the bacteria. Under similar test conditions, the silver-containing alginate fibers had a far better antimicrobial effect than other common types of alginate fibers [43].

4.11 The Production of Other Novel Alginate Fibers

From a wet-spinning point of view, the production process for alginate fibers is one of the cleanest processes used for man-made fibers. The dissolution of sodium alginate takes place in pure water and at neutral pH, while coagulation can take place in a dilute aqueous $CaCl_2$ solution, again at a neutral pH and under room temperature. This makes it easy for alginate to be used as a carrier for biologically active compounds that can be added to the fibers and still maintain their bioactivities in the finished fibers.

Fan et al. [16] prepared fibers from alginate and gelatin blends by spinning their solution through a viscose-type spinneret into a coagulation bath containing aqueous $CaCl_2$ and ethanol. The highest tensile strength was obtained when the gelatin content was 30 wt% of the overall solid content. The water-retention values of the blend fibers increase with the increase in gelatin contents. There was strong interaction and good miscibility between alginate and gelatin molecules, as a result of intermolecular hydrogen bonds.

Wang et al. [62] prepared fibers from blends of alginate and soy protein by spinning their solution through a viscose-type spinneret into a novel coagulation bath containing aqueous $CaCl_2$, HCl, and ethanol. Fibers with 10% soy protein isolate had a tensile strength of 14.1 cN/tex in the dry state and 3.46 cN/tex in the wet state, respectively, with elongation at break of 20.71% and 56.7%, respectively.

Kobayashi et al. [31] used alginate fibers for enzyme immobilization. They spun an aqueous mixture of sodium alginate and enzymes into divalent metallic ion solution as a coagulation bath to produce enzyme-containing fibers. The entrapment yields of enzymes, such as glucoamylase, cyclodextrin glucanotransferase, endo-polygalacturonase, and protease, were higher in the calcium alginate fibers than those found in calcium alginate beads made under similar conditions. It was found that the yields increased with the increase in the extrusion rate through the spinning nozzle because at a higher extrusion rate, the polymeric molecules are more highly oriented along the fiber axis, which can help prevent leakage of the entrapped enzymes.

Polymer blending is a useful tool for the functional modification of alginate fibers and the resultant properties can be tailored for their applications in wound dressings, cosmetic products, baby diapers, adult incontinence products, feminine hygiene products, and other functional medical and hygiene materials where specific functional performances are required. As been pointed out previously, the blending of alginate with carboxymethyl cellulose is proven highly effective in enhancing the gelling properties of alginate fibers whilst the addition of silver-containing particles into alginate fibers produces silver-containing alginate fibers with strong antimicrobial properties. In recent years, there have been many attempts to combine alginate with chitosan through many different techniques in order to generate fibers that have the high absorbency of alginate fibers and strong antimicrobial performance of chitosan, thus possessing two key properties important for medical and healthcare products.

Water soluble derivatives of chitosan such as carboxymethyl chitosan and hydroxypropyl chitosan can be mixed with aqueous sodium alginate solution to prepare a homogeneous spinning solution which can be spun into a polyblend fiber containing both alginate and chitosan. Fan et al. [18] were able to prepare fibers that contain 10–70% carboxymethyl chitosan with antimicrobial properties. When the carboxymethyl chitosan content was 30%, the tensile strength of the polyblend fiber was 13.8 cN/tex [17]. Jiang et al. [28] showed that the addition of carboxymethyl chitosan can significantly increase the absorption properties of alginate fibers. Hu et al.

[26] and Chang et al. [14] used co-electrospinning technique to combine alginate and chitosan in the same polyblend fiber.

Qin [47] developed a novel process whereby alginate and chitosan can be uniformly dispersed in the same spinning solution. In this process, an acid-resistant derivative of alginate, propylene glycol alginate (PGA), was first dissolved in water and then mixed with an acidic chitosan solution. PGA is an esterified derivative of alginate with excellent emulsifying properties, which is esterified with propylene oxide at 70 °C under pressure with alkali as catalyst [37, 57]. Figure 4.10 shows a schematic illustration of the conversion of alginic acid into propylene glycol alginate.

Because the carboxylic acid group is esterified by propylene glycol to yield an ester group, PGA is soluble in water while remaining stable in acidic environment upto pH 3–4, when sodium alginate would precipitate out as alginic acid. This tolerance to acidic environment together with its strong salt resistance makes PGA highly valuable in foodstuffs and beverages with strong acidity and high levels of metallic ions such as calcium and sodium, where PGA is able to improve the stability of acid in foodstuff and also prevent the lay down created by calcium and other high valued metallic ions. In the spinning of the mixed solution of PGA and chitosan, upon extrusion into an alkaline coagulation bath, chitosan precipitates out to form filaments that contain PGA, and during the subsequent processing, part of the ester groups in PGA are hydrolyzed to generate carboxylic acid groups that can hold a large amount of water, with some ester groups reacting with the amine groups in chitosan to form a cross-linked structure that helps to maintain a stable fiber structure.

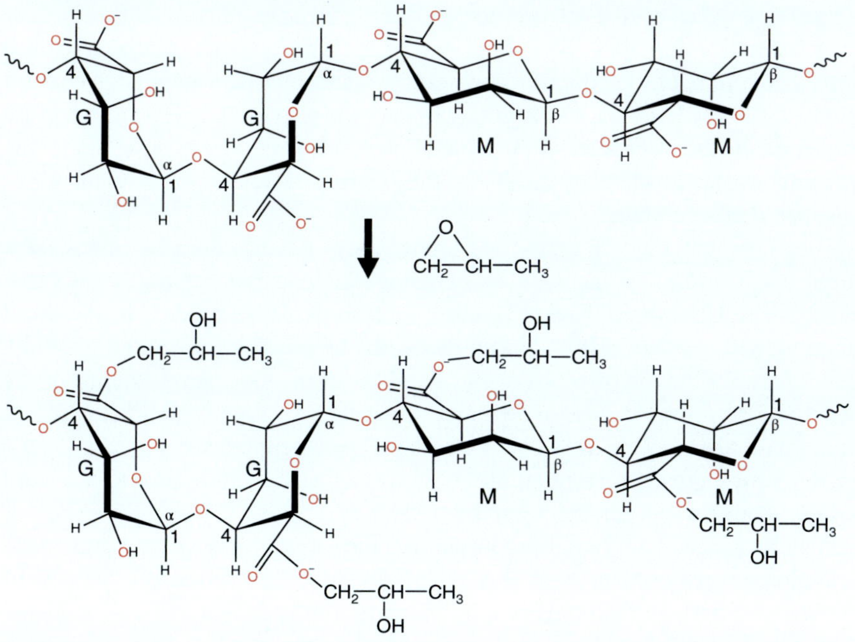

Figure 4.10 Schematic illustration of the conversion of alginic acid into propylene glycol alginate.

Table 4.12 Gel swelling ratio of propylene glycol alginate/chitosan polyblend fibers.

Ratio between PGA and chitosan (g/g)	Gel swelling ratio (g/g)	
	In de-ionized water	In saline
0 : 100	2.91	2.88
5 : 100	3.45	3.10
10 : 100	4.57	4.15
20 : 100	5.78	4.65
30 : 100	7.65	5.22

Table 4.12 shows the gel swelling ratio of propylene glycol alginate/chitosan polyblend fibers made with different proportions of these two polymers. It can be seen that the gel swelling ratio increases with the increased use of propylene glycol alginate.

4.12 Historical Development of Alginate Fibers

Since the first attempt to produce fibers from alginate, there have been many applications of these fibers. In the early stages of development, alginate fibers were used principally in the production of water-soluble yarns that would dissolve in a scouring process. These yarns were used as a support during the manufacture of fine lace, or as draw threads in the production of hosiery. Fabrics from alginate fibers were once produced commercially for their fire-resistant property, because of the high content of metal ions in the fiber [12, 13]. Alginate fibers were also used for the manufacture of bags used for the transportation of soiled hospital linen that was designed to dissolve in the wash. However, by the 1970s, they were replaced for these applications by cheaper synthetic fibers.

The first person in modern times to recognize the potential value of alginate fibers in surgery and wound management was George Blaine, a major in the Royal Army Medical Corps. He showed them to be absorbable in tissue, sterilizable by heat, and compatible with penicillin [8]. He also described how he had used alginate films clotted in situ for the treatment of wounds and burns in troop ship hospitals in the Far East and described the use of alginate, sometimes in combination with plasma as an alginate-plasma film, as "puncture patches" over septal defects.

During a subsequent assessment of the use of alginate as hemostats and wound dressings, Blaine reported their apparent lack of toxicity following a series of animal studies in which fibers were implanted into animal tissues, and gels made from alginate were used to treat experimentally produced burns [9]. Successful use of alginate-derived materials in aural surgery and neurosurgery was reported by Oliver and Blaine [38]. Bray et al. [11] described the results of a three-month trial into the use of alginate in the casualty department of Croydon Hospital where alginates in the form of films, wool, gauzes, and clots formed in situ by mixing sterile solutions of calcium chloride and sodium alginate, were applied to a wide range of wounds, including burns, lacerations, ulcers, and amputations. In all cases, healing was rapid and uneventful. By the late 1940s and early 1950s, alginates were being used in some 70 hospitals in the UK over a range of surgical specialties [10].

The revival of alginate fibers began with a new theory in wound management. Although it has a long history, the standard practice in wound management remained fairly static until Winter published his work in 1962 on acute superficial wounds in the domestic pig [43, 50], which showed that wounds covered with a film dressing maintained a moist environment and healed faster than those left to dry out, heralding the advent of modern wound dressings based on the principle of moist healing [46, 47]. In the early 1980s, as wound dressings based on the moist healing principle were expanding, alginate fibers and nonwoven fabrics were found to have unique gel-forming characteristics whereby on contact with wound exudates, sodium ions in the exudates can exchange with calcium ions in the fibers, and as the ion-exchange process proceeds, the fibers absorb more and more water to form a fibrous gel. For the alginate wound dressings, as water enters the fiber structure, the entire textile structure is transformed into a sheet of moist gel, thus providing an ideal moist healing environment for the underlining wound. Many clinical trials have shown that alginate wound dressings not only have the high absorption capacities but also possess the ability to promote wound healing. In order to further improve its clinical efficacy, many attempts have been made to improve the alginate fibers and wound dressings. Table 4.13 summarized five generations of alginate fibers for the production of wound dressings.

Table 4.13 A summary of five generations of alginate fibers for the production of wound dressings.

Stage	Product	Product characteristics	Main brand
1G	Calcium alginate fiber	Gelling fibers made from high M calcium alginate fibers [63]	Sorbsan™
2G	Calcium sodium alginate fiber	High absorption fibers made from high G calcium sodium alginate fibers [48]	Kaltostat™
3G	Alginate/CMC polyblend fiber	Highly absorbent and gel-forming fibers made from a mixture of alginate and sodium carboxymethyl cellulose [45]	UrgoSorb™
4G	Silver-containing alginate fiber	Antimicrobial alginate fibers and wound dressings containing silver [41, 52]	UrgoSorb Silver™
5G	Alginate and chitosan polyblend fiber	Highly absorbent and antimicrobial fibers and wound dressings that combine the unique properties of alginate and chitosan [44, 49, 58]	GellinS™

4.13 Summary

Alginate is a natural polysaccharide with abundant supply in nature. By processing alginate into fibers, it is possible to obtain novel biomaterials that can be processed further into woven, knitted, nonwoven, and many other forms of composite structures. Because it is a polymeric acid, alginate fibers can be used as a carrier to deliver zinc, silver, and other bioactive metal ions for wound care and other novel applications. In addition, since it is processed in an aqueous solution and in an aqueous coagulation bath at a neutral pH, many bioactive materials such as enzymes can be combined into the alginate fibers to produce enhanced bioactivities.

References

1 Agren, M.S. (1993). Zinc oxide increases degradation of collagen in necrotic wound tissue. *Br. J. Dermatol.* 129 (2): 221–226.
2 Agren, M.S. (1999). Zinc in wound repair. *Arch. Dermatol.* 135 (10): 1273–1274.
3 Agren, M.S. and Mirastschijski, U. (2004). The release of zinc ions from and cytocompatibility of two zinc oxide dressings. *J. Wound Care* 13 (9): 367–369.
4 Annison, G., Cheetham, N.W.H., and Couperwhite, I. (1983). Determination of the uronic acid composition of alginates by high-performance liquid chromatography. *J. Chromatogr.* 204: 137–143.
5 Atkins, E.D.T., Nieduszynski, I.A., Mackie, W. et al. (1973). Structural components of alginic acid. 1. Crystalline structure of poly-β-D-mannuronic acid. Results of x-ray diffraction and polarized infrared studies. *Biopolymers* 12: 1865–1878.
6 Atkins, E.D.T., Nieduszynski, I.A., Mackie, W. et al. (1973). Structural components of alginic acid. 2. Crystalline structure of poly-α-L-guluronic acid. Results of x-ray diffraction and polarized infrared studies. *Biopolymers* 12: 1879–1887.
7 Bixler, H.J. and Porse, H. (2011). A decade of change in the seaweed hydrocolloids industry. *J. Appl. Phycol.* 23 (3): 321–335.
8 Blaine, G. (1946). The use of plastics in surgery. *Lancet* 1946: 525–552.
9 Blaine, G. (1947). Experimental observations on absorbable alginate products in surgery. *Ann. Surg.* 125 (1): 102–114.
10 Blaine, G. (1951). A comparative evaluation of absorbable haemostatics. *Postgrad. Med. J.* 27: 613–620.
11 Bray, C., Blaine, G., and Hudson, P. (1948). New treatment for burns, wounds and haemorrhage. *Nurs. Mirror* 86: 239–242.
12 Chamberlain, N.H., Johnson, A., and Speakman, J.B. (1945). Some properties of alginate rayons. *J. Soc. Dye. Colour.* 61 (1): 13–20.
13 Chamberlain, N.H., Lucas, F., and Speakman, J.B. (1949). The action of light on calcium alginate rayon. *J. Soc. Dye. Colour.* 65 (12): 682–692.
14 Chang, J., Lee, Y.H., Wu, M.H. et al. (2012). Electrospun anti-adhesion barrier made of chitosan alginate for reducing peritoneal adhesions. *Carbohydr. Polym.* 88: 1304–1312.

15 Dealey, C. (1994). *The Care of Wounds*. Oxford: Blackwell Science Ltd.

16 Fan, L., Du, Y., Huang, R. et al. (2005). Preparation and characterization of alginate/gelatin blend fibers. *J. Appl. Polym. Sci.* 96 (5): 1625–1629.

17 Fan, L., Du, Y., Zhang, B. et al. (2005). Antibacterial fibers made of calcium alginate/chitosan derivative composite. *J. Funct. Polym.* 18 (3): 34–38.

18 Fan, L., Du, Y., Zhang, B. et al. (2006). Preparation and properties of alginate/carboxymethyl chitosan blend fibers. *Carbohydr. Polym.* 65: 447–452.

19 Furr, J.R., Russell, A.D., Turner, T.D. et al. (1994). Antibacterial activity of Actisorb Plus, Actisorb and silver nitrate. *J. Hosp. Infect.* 27 (3): 201–208.

20 Grant, G.T., Morris, E.R., Rees, D.A. et al. (1973). Biological interactions between polysaccharides and divalent cations: the egg-box model. *FEBS Lett.* 32: 195–198.

21 Grasdalen, H., Larsen, B., and Smidsrod, O. (1979). Study of the composition and sequence of uronate residues in alginates. *Carbohydr. Res.* 68: 23–31.

22 Haug, A. and Larsen, B. (1962). Quantitative determination of the uronic acid composition of alginates. *Acta Chem. Scand.* 16: 1908–1918.

23 Haug, A., Larsen, B., and Smidsrod, O. (1966). A study of the constitution of alginic acid by partial acid hydrolysis. *Acta Chem. Scand.* 20: 183–190.

24 Haug, A., Larsen, B., and Smidsrod, O. (1967). Correlation between chemical structure and physical properties. *Acta Chem. Scand.* 21: 768–778.

25 Haug, A., Larsen, B., and Smidsrod, O. (1974). Uronic acid sequence in alginate from different sources. *Carbohydr. Res.* 32: 217–225.

26 Hu, W. and Yu, H. (2013). Coelectrospinning of chitosan/alginate fibers by dual-jet system for modulating material surfaces. *Carbohydr. Polym.* 95: 716–727.

27 Ji, M.H. (1984). Studies on the M:G ratios in alginate. *Hydrobiologia* 116 (117): 554–556.

28 Jiang, L., Kong, Q., Ji, Q. et al. (2007). Study on preparation and properties of sodium alginate/water-soluble chitin blend fiber. *Synth. Fibres China* (12): 12–15.

29 Klasen, H.J. (2000). Historical review of the use of silver in the treatment of burns. I. Early uses. *Burns* 26 (2): 117–130.

30 Klasen, H.J. (2000). Historical review of the use of silver in the treatment of burns. II. Renewed interest for silver. *Burns* 26 (2): 131–138.

31 Kobayashi, Y., Matsuo, R., Ohya, T. et al. (1987). Enzyme-entrapping behaviors in alginate fibers and their papers. *Biotechnol. Bioeng.* 30 (3): 451–457.

32 Kobayashi, Y., Kamishima, H., Fukuoka, S., et al. (1995). Water soluble alginate fibers. US Patent 5,474,781.

33 Lansdown, A.B.G. (2002). Silver 2: toxicity in mammals and how its products aid wound repair. *J. Wound Care* 11 (5): 173–177.

34 Lansdown, A.B.G., Jensen, K., and Jensen, M.Q. (2003). Contreet hydrocolloid and Contreet foam: an insight into new silver-containing dressings. *J. Wound Care* 12 (6): 205–210.

35 Le, Y., Anand, S.C., and Horrocks, A.R. (1997). Using alginate fiber as drug carrier for wound dressing. In: *Medical Textiles'96* (ed. S.C. Anand), 21–26. Cambridge: Woodhead Publishing Ltd.

36 McDowell, R.H. (1960). Applications of alginates. *Rev. Pure Appl. Chem.* 10: 1–19.

37 Noto, V.H. and Pettitt, D.J. (1980). Production of propylene glycol alginic acid esters. British Patent 1563019.

38 Oliver, L.C. and Blaine, G. (1950). Haemostasis with absorbable alginates in neurosurgical practice. *Br. J. Surg.* 37: 307–310.

39 Onsoyen, E. (1992). Alginates. In: *Thickening and Gelling Agents for Food* (ed. A. Imeson), 1–24. Glasgow: Blackie Academic and Professional.

40 Penman, A. and Sanderson, G.R. (1972). A method for the determination of uronic acid sequence in alginates. *Carbohydr. Res.* 25: 280–285.

41 Qin, Y. (2005). Calcium sodium alginate fibers. *Chem. Fibers Int.* 55 (2): 98–99.

42 Qin, Y. (2005). The ion exchange properties of alginate fibers. *Text. Res. J.* 75 (2): 165–168.

43 Qin, Y. (2005). Silver containing alginate fibers and dressings. *Int. Wound J.* 2 (2): 172–176.

44 Qin, Y. (2006). The characterization of alginate wound dressings with different fiber and textile structures. *J. Appl. Polym. Sci.* 100 (3): 2516–2520.

45 Qin, Y. (2008). Alginate fibres: an overview of the production processes and applications in wound management. *Polym. Int.* 57 (2): 171–180.

46 Qin, Y. (2014). Preparation method and applications of chitosan fibers modified by oxidized alginate. Chinese Patent Zl201310127978.X.

47 Qin, Y. (2015). Preparation methods and applications of polyblend fibers from chitosan and propylene glycol alginate. Chinese Patent Zl201210553968.8.

48 Qin, Y. (2016). *Medical Textile Materials*, 89–107. Cambridge: Elsevier.

49 Qin, Y. and Gilding, D.K. (2000). Fibers of Cospun alginates, US Patent 6,080,420.

50 Qin, Y. and Groocock, M.R. (2007). Polysaccharide fibers. US Patent 7,229,689 B2.

51 Qin, Y., Agboh, C., Wang, X. et al. (1996). Alginate fibers. *Chem. Fibers Int.* 46: 272–273.

52 Qin, Y., Hu, H., and Luo, A. (2006). The conversion of calcium alginate fibers into alginic acid fibers and sodium alginate fibers. *J. Appl. Polym. Sci.* 101 (6): 4216–4221.

53 Smidsrod, O. and Haug, A. (1972). Dependence upon the gel-sol state of the ion-exchange properties of alginates. *Acta Chem. Scand.* 26: 2063–2074.

54 Smidsrod, O., Haug, A., and Whittington, S.G. (1972). The molecular basis for some physical properties of polyuronides. *Acta Chem. Scand.* 26: 2563–2564.

55 Speakman, J.B. and Chamberlain, N.H. (1944). The production of rayon from alginic acid. *J. Soc. Dye. Colour.* 60: 264–272.

56 Stanford E C (1881). Improvements in the manufacture of useful products from seaweeds. British Patent 142.

57 Steiner, A.B. (1947). Manufacture of glycol alginates. US Patent 2426215.

58 Stevens, J. and Chaloner, D. (2005). Urgosorb dressing: management of acute and chronic wounds. *Br. J. Nurs.* 14 (15): S22–S28.

59 Thomas, S. and McCubbin, P. (2003). A comparison of the antimicrobial effects of four silver-containing dressings on three organisms. *J. Wound Care* 12 (3): 101–107.

60 Thomas, S. and McCubbin, P. (2003). An *in vitro* analysis of the antimicrobial properties of 10 silver-containing dressings. *J. Wound Care* 12 (8): 305–308.

61 Thomas, A., Gilson, C.D., and Ahmed, T. (1995). Gelling of alginate fibers. *J. Chem. Technol. Biotechnol.* 64: 73–77.

62 Wang, Q., Du, Y., Hu, X. et al. (2006). Preparation of alginate/soy protein isolate blend fibers through a novel coagulating bath. *J. Appl. Polym. Sci.* 101 (1): 425–431.

63 Winter, G.D. (1962). Formation of the scab and the rate of epithelialization of superficial wounds in the skin of the young domestic pig. *Nature* 193: 293–294.

64 Woodward, F. (1951). The Scottish seaweed research association. *J. Mar. Biol. Assoc. U. K.* 29 (3): 719–725.

5

Ion-Exchange and Gel-Forming Properties of Alginate Fibers

5.1 Introduction

Alginate fibers have had a long history. As early as 1944, Speakman and Chamberlain [19] reported a detailed process for the production of alginate fibers and in the following years, more studies were carried out for the applications of these fibers in the textile industry [1, 2, 4]. In the 1950s, alginate fibers were used as draw threads for sock production and as home decoration materials, utilizing their ability to dissolve in a dilute aqueous alkali solution and their inflammable properties.

Since the 1980s, alginate fibers have been widely used in the production of functional wound dressings. In this particular application, alginate fibers have unique ion-exchange and gel-forming characteristics whereby on contact with wound exudates, sodium ions in the exudates exchange with calcium ions in the fiber, and as more and more sodium ions enter the fiber structure, the resultant calcium sodium alginate fibers absorb more and more water to become a fibrous gel [10, 15]. When used on wound, this in situ formed hydrogel provides an ideal moist healing environment that can promote wound healing [11, 12].

This chapter summarizes the ion-exchange properties between alginate fibers and a number of metal ions and describes the gelling characteristics of several types of alginate fibers.

5.2 Characterization Methods for Ion Exchange and Gel Forming Properties

Since their introduction into the wound management market, many types of commercially available wound dressings made from different types of alginate fibers have been marketed globally by many producers and distributors. Table 5.1 summarizes the key structural features of 7 types of commercially available alginate wound dressings which are used in the studies described in this chapter [10].

The commercially available alginate wound dressings in Table 5.1 represent a wide variety of chemical and physical structures of alginate fibers, which were used to assess the effect of structural parameters on product performances. Among the three

Alginate Fibers and Wound Dressings: Seaweed Derived Natural Therapy, First Edition. Yimin Qin.

Table 5.1 Key structural features of seven types of commercially available alginate wound dressings.

Brand name of alginate dressing	Guluronate and mannuronate content	Calcium and sodium content	Nonwoven structure
Sorbsan	~39/61	~96.6/3.4	Un-needled
Tegagel™	~39/61	~96.6/3.4	Hydro-entangled
Algosteril	~68/32	~99.6/0.4	Needled
Kaltostat	~68/32	~80/20	Needled
Tegagen HG™	~38/62	~65/35	Needled
Curasorb™	~68/32	~99.2/0.8	Needled
Urgosorb™ (Alginate/CMC blend)	~68/32	~95.2/4.8	Needled

main products, Sorbsan™ is made of calcium alginate fiber in a loose nonwoven format, Algosteril™ is a needled felt of calcium alginate, and Kaltostat™ is made of calcium/sodium alginate fiber in a needled nonwoven format. These products were made from alginate containing different proportions of α-L-guluronic acid and β-D-mannuronic acid. After chemical treatment to the fibers, they also contain different amount of calcium and sodium ions.

The calcium and sodium contents are expressed as the percentage of carboxylic acid groups being as calcium salt or sodium salt. This assumes that all the carboxylic acid groups on the alginate molecular chain exist either as calcium salt or as sodium salt. To measure the calcium and sodium contents, the fibers were first digested in 98% sulfuric acid solution, and the calcium and sodium contents were measured with atomic absorption spectrometer. Assuming the wt/wt calcium content of the fiber is C_1 and the sodium content is C_2, the fiber calcium content equals to $[C_1/20]/[C_1/20 + C_2/23] \times 100\%$, while the fiber sodium content is $[C_2/23]/[C_1/20 + C_2/23] \times 100\%$ (note: one calcium ion binds with two carboxylic acid group).

The gel swelling abilities of the alginate fibers are measured by placing 0.2 g fiber in 100 ml of either distilled water or 0.9% wt/wt aqueous sodium chloride solution (normal saline). After one hour, the fibers are separated with the contacting solution and placed in a centrifuge tube with the bottom half filled with knitted viscose rayon fabric to contain the spin-off solution. The centrifuge is carried out at 1200 rev/min for 15 minutes. After that, the fiber (W_1) is dried at 105 °C to constant weight (W_2). The gel swelling ratio is expressed as the ratio between the weight of the wet sample and that of the dry sample, i.e. W_1/W_2.

In order to measure the gel strength, 2 g of alginate fibers was first cut to about 5 mm length. They were placed in a beaker with 50 ml distilled water. After stirring for a few minutes, 10 ml of 9% wt/wt aqueous tri-sodium citrate solution is added into the above mixture. The resultant mixture contains about 3% alginate and 1.5%

tri-sodium citrate. After thorough mixing, the mixture is left for five minutes before measuring gel strength on a Stevens gel strength meter.

The ion exchange of alginate fibers with the contacting solution is conducted with solution A as the contacting medium. The British Pharmacopoeia specifies solution A as an aqueous solution containing 142 mmol of sodium chloride and 2.5 mmol of calcium chloride. In total, 1 g fiber is placed in 40 ml of solution A and is conditioned at 37 °C for 30 minutes. After separating the fiber from the solution, the calcium and sodium contents in the solution were measured by atomic absorption spectrometer. If the calcium content in the original solution is C_1 ppm, and that after the ion exchange process is C_2 ppm, then the calcium ions released per gram of fiber equal to $40 \times (C_2 - C_1) \times 10^{-6}$ g/g.

The absorption capacities of the dressings were measured according to the method specified in the British Pharmacopoeia monograph for alginate dressings and packings. A piece of 5 cm × 5 cm dressing (W_1) is placed in a flat-bottomed Petri dish 90 mm in diameter, to the dish added a quantity of solution A 40 times the weight of the dressing. After conditioning at 37 °C for 30 minutes, the sample is lifted from one corner and held in the air for 30 s before the weight of the sample is measured (W_2). The absorption capacity is expressed as $(W_2 - W_1)/W_1$ g/g. Figure 5.1 shows an illustration of the testing procedure

The changes in fiber length after wetting were measured under an optical microscope. A piece of fiber about 2 mm in length is wetted with either distilled water or 0.9% saline, and the length change is expressed as (wet length – dry length)/dry length × 100%.

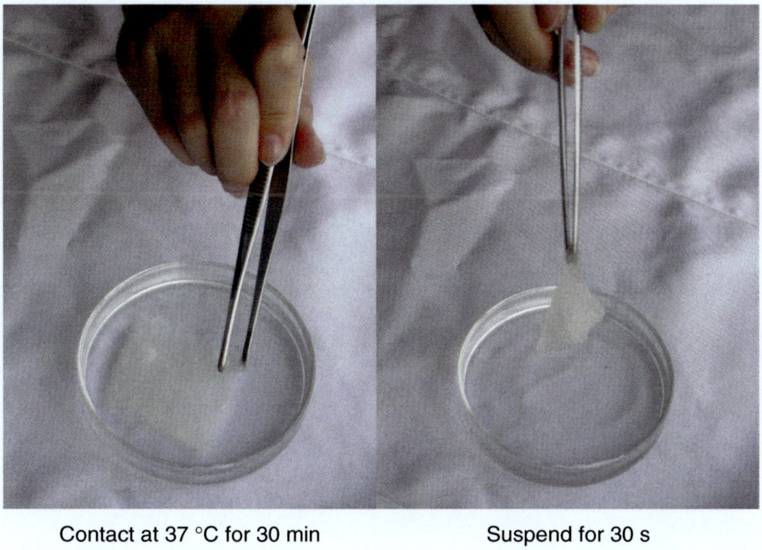

Figure 5.1 An illustration of the absorption test procedure for alginate wound dressing.

5.3 Ion-Exchange Properties of Alginate Fibers

5.3.1 Ion-Exchange Between Calcium Alginate Fibers and Sodium Ions

Table 5.2 shows the concentrations of calcium ions in the contact solution after alginate fibers were in contact with solution A. It is clear that the amount of calcium ions released was significantly different for Algosteril and Sorbsan. In the solution containing Algosteril, 321.5 ppm of calcium ions were measured, while for Sorbsan, the concentration was 557.5 ppm, almost twice that of the Algosteril sample. The release of calcium ions from the Sorbsan sample is equivalent to 1.86% of the fiber weight.

It is known that the Algosteril fibers are made from a grade of alginate rich in guluronic acid, while the Sorbsan fibers are made with alginate containing a high percentage of mannuronic acid. Literature studies have shown that guluronic acid has a stronger binding to calcium ions than mannuronic acid [3, 8, 17, 18]. Chemically, when calcium ions form salt bonds with alginic acid, they fit into the space between two neighboring guluronic acid units, creating what is called an egg box structure, which is highly stable and difficult to be replaced by sodium ions. In this sense, alginate fibers with a high percentage of guluronic acid, or high G alginate, have stronger binding with calcium ions and are therefore more difficult to exchange with sodium ions when in contact with saline solution.

The ion exchange between calcium ions in the alginate fibers and wound dressing and sodium ions in the wound exudate has a practical significance in that when part of the calcium ions in the fibers are replaced by sodium ions in the solution, water is drawn into the fiber since sodium alginate is water soluble. As more and more water is drawn into the fiber, it swells to become a gel that has fluid retention ability and hence helps to keep a moist healing environment on the wound surface. Figure 5.2

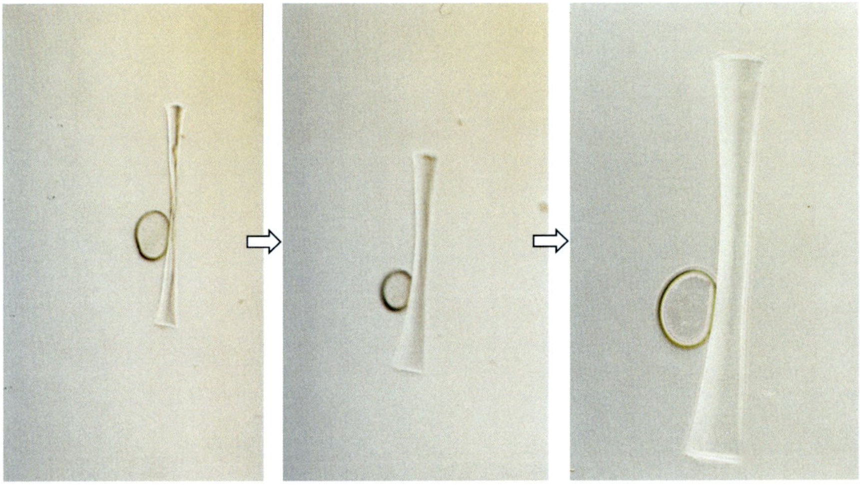

Figure 5.2 An illustration of the gelling process when calcium alginate fiber is in contact with saline solution.

Table 5.2 Release of calcium ions from alginate fibers when in contact with sodium ions.

	Test sample		
Test criteria	**Solution A**	**Algosteril**	**Sorbsan**
Ca(II) concentration in the contact solution (ppm)	92.5	321.5	557.5
Calcium release as % of fiber weight	N/A	0.91%	1.86%

shows the process of gel formation when a piece of calcium alginate fiber is placed in contact with saline solution.

As Table 5.2 clearly shows, it is far more difficult for the high G-type Algosteril fibers to exchange with sodium ions, giving them poorer gelling ability than the high M-type Sorbsan fibers. One way to improve the gelling ability of high G alginate fibers is to introduce sodium ions into the fibers before being applied to the wound surface. In order to make fibers with different levels of sodium ions, the original calcium alginate fibers can be treated with aqueous solutions containing different amount of Na_2SO_4, which is able to replace part of the calcium ions in the fibers with sodium ions, since the reaction between calcium alginate fiber and Na_2SO_4 results in the formation of $CaSO_4$ which is only slightly soluble in water.

As previously mentioned in Table 4.5, after the calcium alginate fibers were treated with aqueous solutions containing various levels of Na_2SO_4, it was found that as the Na_2SO_4 concentration increased, more and more calcium ions were replaced by sodium ions. Gel swelling results show that while the original sample swelled only a little in water, with a gel swelling ratio of 1.8, there were significant increases in the gel swelling ratio as more and more calcium alginate was converted into sodium alginate. The fiber treated with 0.7% Na_2SO_4 has a gel swelling ratio of 21 [8].

5.3.2 Ion-Exchange Between Alginate Fibers and Zinc Ions

Zinc is an important microelement in the human body, playing a unique role in the wound healing process. Similar to calcium ions, zinc ions can be used in the coagulation bath to precipitate sodium alginate, resulting in the formation of zinc alginate fibers. A study by Mikołajczyk and Wołowska-Czapnik [5] showed that under appropriate conditions, the tenacity of zinc alginate fibers can reach 28.65 cN/tex, where the zinc ion content in the fibers is 164.4 mg/g.

In order to prepare zinc alginate fibers containing different amount of zinc ions, calcium alginate fibers can be treated with zinc salt to load the fibers with different amount of zinc ions [14, 16]. During the preparation process, calcium alginate fibers were first treated with aqueous hydrochloric acid solution to convert the fibers into alginic acid, before placing them in contact with aqueous zinc sulfate solution. Table 5.3 shows the effect of treatment time on the zinc ion content of alginate fibers. After 0.5 hour, the zinc content in the fiber was 116.0 mg/g, while after 24 hours, zinc content was 122.6 mg/g, which means that the zinc content at 0.5 hour is 94.6% of

Table 5.3 Effect of treatment time on the zinc ion content of alginate fibers.

Treatment time (h)	Dry weight of fiber (g)	Zinc concentration in the digest solution (mg/l)	Zinc content in the fiber (mg/g)
0.5	0.938	5.44	116.0
1	1.054	6.32	119.9
3	0.936	5.65	120.7
8	0.924	5.61	121.4
24	0.687	4.21	122.6

Table 5.4 Effect of $ZnSO_47H_2O$ concentration on the zinc ion content of alginate fibers.

Amount of $ZnSO_47H_2O$ (g)	Amount of alginic acid fiber (g)	Zinc content in the digest solution (mg/l)	Zinc content in the fiber (mg/g)
0.5	0.595	1.58	53.1
1	0.873	3.07	70.3
3	0.521	2.84	109.0
10	0.839	5.39	128.5

that at 24 hours, showing that the absorption of zinc ions by alginic acid fibers is a fairly fast process [14].

Table 5.4 shows the effect of $ZnSO_47H_2O$ concentration on the absorption of zinc ions by alginic acid fiber. When excess amount of $ZnSO_47H_2O$ was used, the fiber zinc content reached 128.5 mg/g, which compares to zinc content of 164.4 mg/g for zinc alginate fibers obtained through the wet spinning process using aqueous zinc chloride as the coagulation bath. These two sets of results show that when alginic acid fibers were treated with $ZnSO_47H_2O$, the majority of the carboxylic acid groups were able to combine with zinc ions to form zinc alginate. As illustrated in Table 5.4, by controlling the ratio between the fiber and $ZnSO_47H_2O$, it was possible to control the zinc content of the alginate fibers [14].

Table 5.5 shows the effect of temperature on the zinc ion content of alginate fibers when alginic acid fibers were treated with $ZnSO_4$. When the temperature was 30, 40, 50, 60, and 80 °C, the zinc content in the fiber was 103.0, 109.5, 110.8, 119.8, and 127.9 mg/g, respectively, indicating that higher treatment temperature benefits the absorption of zinc ions. When comparing the absorption of zinc ions by high M and high G alginic acid fibers, results showed that the zinc contents were 127.9 and 132.5 mg/g, respectively, in the high M fiber and high G fiber, indicating that high G alginic acid fiber is better at absorbing zinc ions [14].

Figure 5.3 shows the zinc ion concentration in the contact solution when zinc alginate fibers were placed in contact with solution A at different temperatures.

Table 5.5 Effect of treatment temperature on the zinc ion content of alginate fibers.

Treatment temperature (°C)	Dry weight of fiber (g)	Zinc content in the digest solution (mg/l)	Zinc content in the fiber (mg/g)
30	0.829	4.27	103.0
40	0.869	4.76	109.5
50	0.819	4.54	110.8
60	1.135	6.8	119.8
80	0.483	3.09	127.9

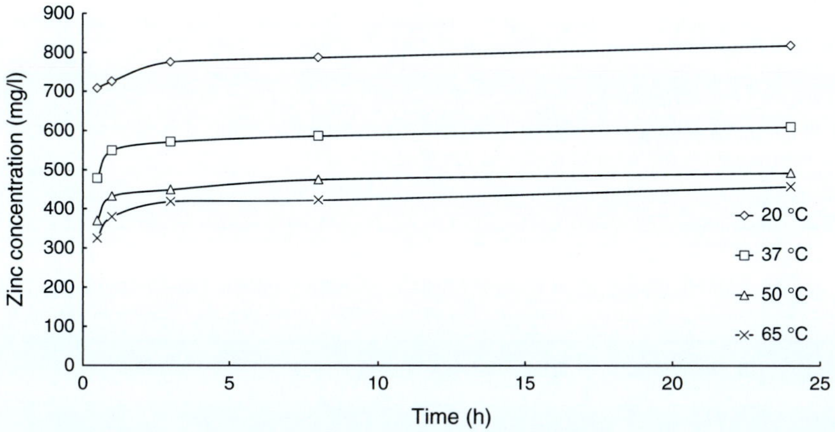

Figure 5.3 Zinc ion concentration in the contacting solution when zinc alginate fibers were placed in contact with solution A at different temperatures. Source: Adapted with permission from Qin and Chen [14].

It can be seen that the release of zinc ions from zinc alginate fiber is a fairly fast process, with the zinc ion concentration in the contact solution reaching equilibrium in three hours. It is worth noting that the fibers release more zinc ions at lower temperatures. After being in contact for 0.5 hour, the zinc concentration was 708 mg/l at 20 °C, while at 65 °C, the zinc concentration was 324 mg/l over the same period. After 24 hours, the zinc concentrations in the contact solutions were 818 and 457 mg/l at 20 and 65 °C, respectively.

Figure 5.4 shows the zinc ion concentration in the contact solution when zinc alginate fibers were placed in contact with aqueous solutions containing different amount of protein at 37 °C. After 0.5 hours, the zinc ion concentrations were 92, 347, and 626 mg/l in the contact solutions containing 1.0%, 2.9%, and 5.0% protein, respectively, showing the chelating effect of protein in promoting the release of zinc ions.

The ion exchange between alginate fibers and zinc ions can be used to prepare fibers with different zinc contents [16]. As shown in Table 5.6, in three separate

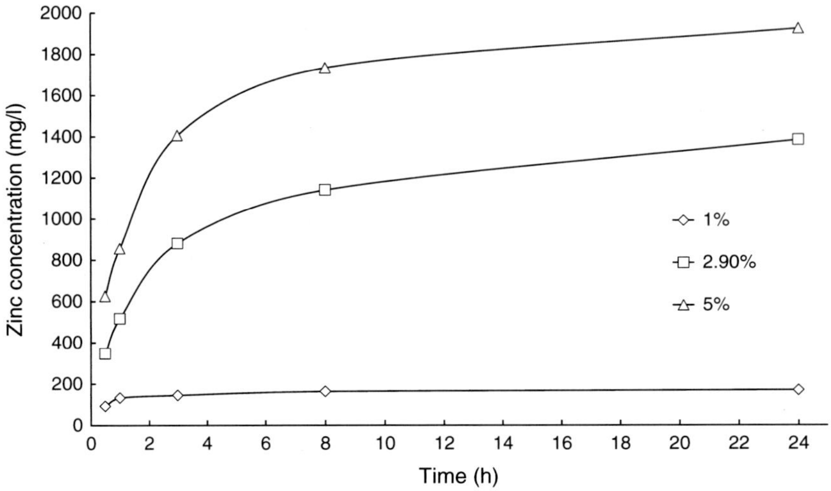

Figure 5.4 Zinc ion concentration in the contact solution when zinc alginate fibers were placed in contact with aqueous solutions containing different amount of protein at 37 °C. Source: Adapted with permission from Qin and Chen [14].

Table 5.6 Calcium and zinc ion contents in $Zn_2SO_47H_2O$ treated calcium alginate fibers.

Sample no.	Amount of $Zn_2SO_47H_2O$ (g)	Amount of calcium alginate fiber (g)	Calcium content in the fiber (mg/g)	Zinc content in the fiber (mg/g)
1	0.774	5.192	166.4	46.5
2	2.331	5.216	127.4	84.9
3	7.885	5.292	98.9	124.6
4[a)]	N/A	N/A	23.3	164.4

a) Zinc alginate fiber made in the wet spinning process with zinc chloride as the coagulant.

samples, 0.774, 2.331, and 7.885 g of $Zn_2SO_47H_2O$ were dissolved in 1 l de-ionized water before being placed in contact with 5.192, 5.216, and 5.292 g calcium alginate fibers, respectively. After 24 hours, the fibers were washed and dried before being tested for zinc and calcium contents. These three samples have zinc contents of 46.5, 84.9, and 124.6 mg/g, respectively, which means that by controlling the ratio between calcium alginate fibers and $Zn_2SO_47H_2O$ in the aqueous solution, it is possible to obtain fibers with different levels of zinc contents. In comparison to the zinc-containing alginate fibers obtained through the indirect ion-exchange method, zinc alginate fiber made in the wet-spinning process with zinc chloride as the coagulant contained 23.3 and 164.4 mg/g calcium and zinc ions, respectively.

Table 5.7 shows the antimicrobial effect against *Bacillus subtilis var. nier* of zinc alginate fibers with different zinc ion contents. Pure zinc alginate fibers made from the wet-spinning process showed a reduction rate of 99.77% against *B. subtilis var. nier*, while the fibers with zinc contents of 124.6 and 84.9 mg/g had reduction rates

Table 5.7 The antimicrobial effect against *Bacillus subtilis var. nier* of zinc alginate fibers with different zinc ion contents.

Fiber zinc content (mg/g)	Bioburden at the beginning	Bioburden after 24 h	Reduction of bacteria count (%)
46.5	1.1×10^5	1.1×10^5	0
84.9	1.3×10^5	1.2×10^5	7.69
124.6	1.1×10^5	7.9×10^4	28.18
164.4	5.3×10^4	1.2×10^2	99.77

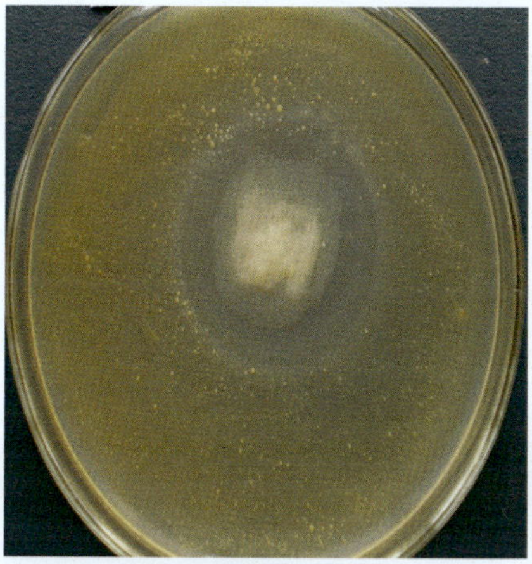

(1) Against *Staphylococcus aureus*

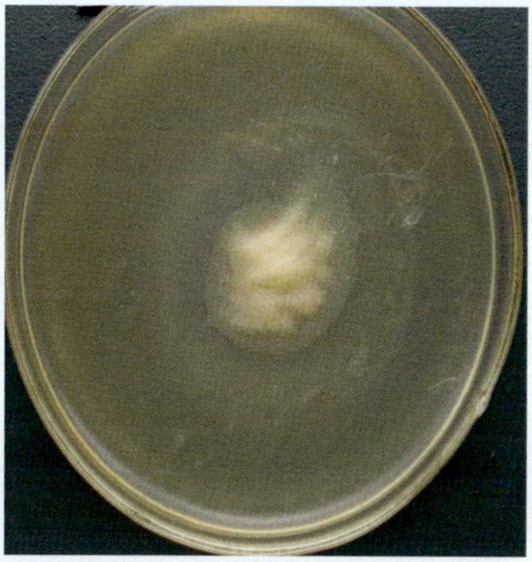

(2) Against *Escherichia coli*

Figure 5.5 Zone of inhibition for zinc alginate fibers. Source: Adapted with permission from Qin et al. [16].

of 28.18% and 7.69%, respectively, with the fibers containing 46.5 mg/g zinc showing no antimicrobial effect [16].

Figure 5.5 shows the zone of inhibition for zinc alginate fibers against *Staphylococcus aureus* and *Escherichia coli*. It is clear that with the release of zinc ions, zinc alginate fibers have zones of inhibition against the two bacteria commonly encountered during wound healing. The antimicrobial property is further illustrated in Figure 5.6, which demonstrated that under the same test conditions, calcium alginate fibers were unable to stop the spreading of bacteria, while zinc alginate fibers were able to inhibit the migration of bacteria through the nonwoven structure.

5.3.3 Ion-Exchange Between Alginate Fibers and Copper Ions

As a polymeric acid, alginate can be in the form of alginic acid, sodium alginate, calcium alginate, and many other salts. Alginic acid fibers are produced when calcium ions are removed by treating calcium alginate fibers with aqueous acid solutions. Calcium/sodium alginate fibers can be obtained by replacing part of the calcium

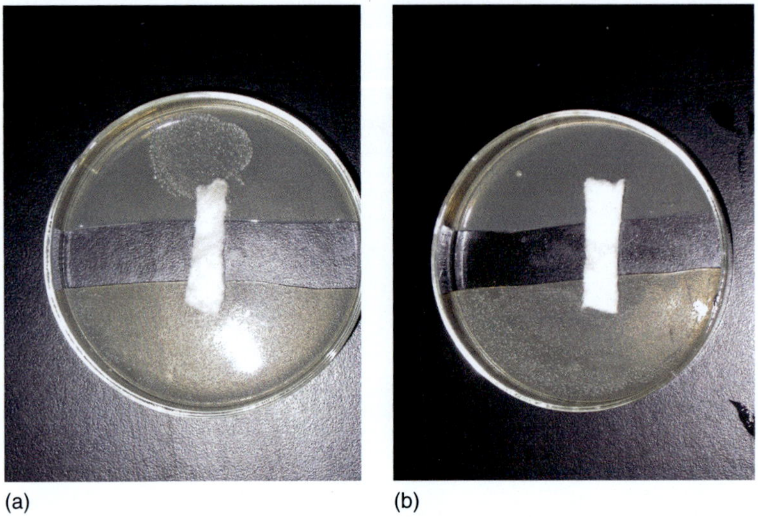

Figure 5.6 Antimicrobial effect of alginate fibers, (a) calcium alginate fibers; (b) zinc alginate fibers. Source: Adapted with permission from Qin et al. [16].

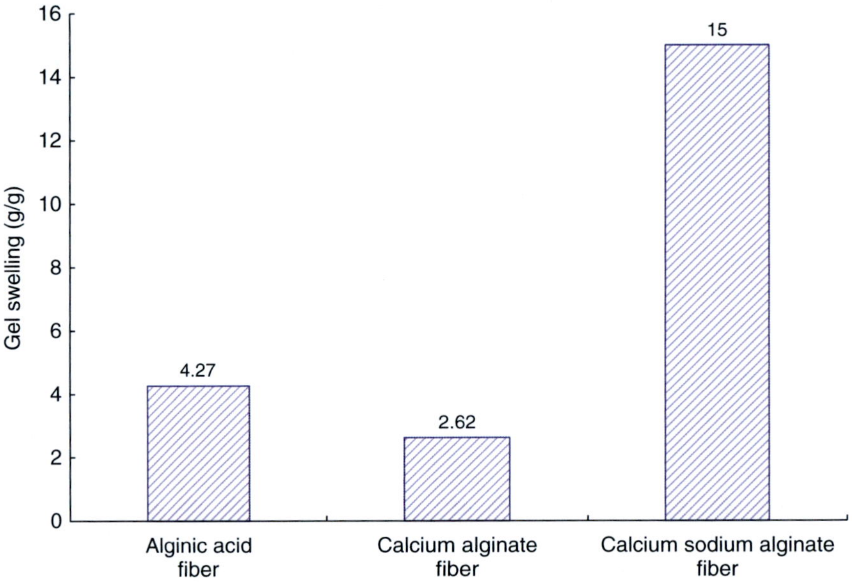

Figure 5.7 The swelling ratio of three types of alginate fibers when wet in water. Source: Adapted with permission from Mo et al. [6].

ions in the fibers with sodium ions. When alginic acid fiber, calcium alginate fiber, and calcium/sodium alginate fiber were placed in water, they showed significantly different water absorption behavior, with alginic acid fiber absorbing more water than calcium alginate fiber, and calcium/sodium alginate fiber showing the highest absorption capacity. Figure 5.7 shows the swelling ratio of the three types of fibers when wet in water.

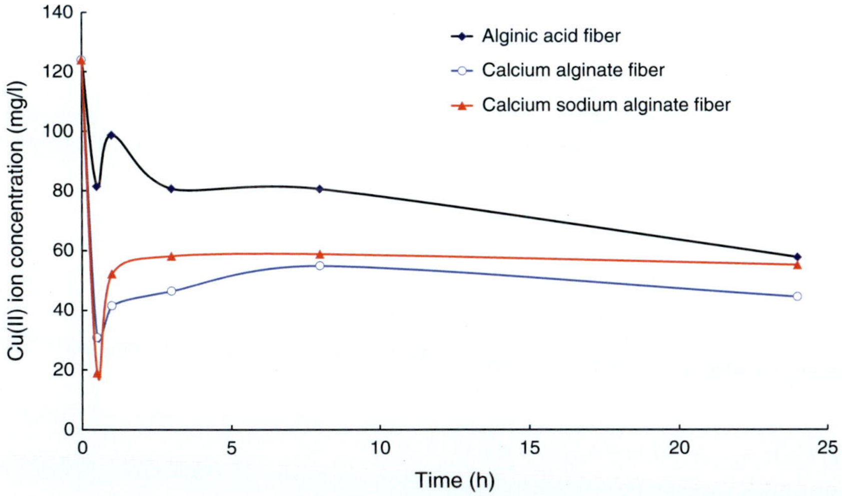

Figure 5.8 The variation in copper ion concentration after the addition of three types of alginate fibers into aqueous copper solution. Source: Adapted with permission from Mo et al. [6].

Figure 5.8 shows the variation in copper ion concentration after the addition of three types of alginate fibers into aqueous copper solution. Among divalent metal ions, copper has one of the strongest bindings with alginic acid and it is clear that the copper ion concentration dropped sharply when the three types of alginate fibers were in contact with the copper solution. Calcium/sodium alginate fiber is capable of a high degree of swelling, hence it is easy for copper ions to diffuse into the fiber structure, thus allowing a fast absorption of copper ions. As the ion exchange proceeds, the copper ion concentration in the solution gradually reached equilibrium, with the three types of fibers showing different levels of copper absorption.

With the three types of alginate fibers used in Figure 5.8, there is a general trend that when in contact with copper solution, the copper ion concentration dropped before recovering to equilibrium. Using calcium alginate fiber as an example, the chemical reaction during the ion exchange process is as follows:

$$\text{Calcium alginate} + \text{Copper sulfate} = \text{Copper alginate} + \text{Calcium sulfate}$$

Since the ion exchange occurs between copper ions in the solution and calcium alginate fiber as a solid, at the beginning of the process, a large amount of copper ions are drawn into the fiber structure to form copper alginate and at the same time, calcium sulfate is formed within the fiber, causing a rapid drop in solution copper concentration. As calcium salt then slowly diffuses out of the fiber, it pulls some of the copper ions back into the solution, which increases the solution's copper concentration.

Table 5.8 shows the absorption capacities of copper ions by the three types of alginate fiber. After 24 hours, the equilibrium absorption of copper ions for alginic acid fiber, calcium alginate fiber, and calcium/sodium alginate fibers was 68.6, 81.7, and 71.0 mg/g, with calcium alginate fibers showing the highest absorption for coppers. This is probably due to the fact that with calcium alginate fibers, the by-product in

Table 5.8 The absorption capacities for copper ions by three types of alginate fibers.

	Absorption of copper ions (mg/g)		
Contact time (h)	**Alginic acid fiber**	**Calcium alginate fiber**	**Calcium/sodium alginate fiber**
0.5	42.5	93.5	105.0
1	26.2	83.2	72.3
3	44.9	79.0	67.0
8	45.8	71.3	66.9
24	68.6	81.7	71.0

the ion exchange is a slightly water-soluble calcium sulfate, unlike sulfuric acid and sodium sulfate formed with other two types of fibers [6].

5.4 Gelling Properties of Alginate Fibers

Alginate is known as an excellent gelling agent whereby the treatment of an aqueous sodium alginate solution with a divalent cation such as Ca^{2+} results in the formation of an ionotropic gel, in which the ion-binding ability of alginate is the basic feature controlling hydrogel formation. Overall, an alginate gel can be depicted as a continuous network swollen with water where physical cross-links, represented by the ion-induced junction zones, hold together different polysaccharide chains. The formation of the hydrogel is favored by the fact that Ca^{2+}–alginate gels have a markedly higher affinity than Na^{+}–alginate solution toward Ca^{2+} ions. This has been explained theoretically as a near-neighbor autocooperative process that predicts that affinity toward a specific ion increases with increasing content of the same ion in the medium. The gelling kinetics is strongly dependent on the method used for the introduction of the cross-linking ion and hence homogeneous or inhomogeneous hydrogel can be obtained.

Since alginate is a copolymer of guluronic acid (G) and mannuronic acid (M), and since these two isomers have much different stereochemical structures, the G and M contents of the fiber have a significant effect on the gelling abilities of the alginate fibers. The GG block is able to form a space between the two monomer units which is ideal for the calcium ion. One calcium ion can form a very strong cross-link between two neighboring GG blocks. For this reason, alginate high in G content can form strong and firm gels while those rich in M contents tend to form weak and soft gels, since the MM block adopts a flat structure and its ability to bind calcium ions is low.

During the wet-spinning process, sodium alginate is extruded into aqueous calcium chloride solution to form swollen calcium alginate fiber, which is then dried to form fibers which are chemically calcium alginate. On contact with wound exudate or other liquid containing sodium ions, alginate fibers have a unique ion exchange property whereby the calcium ions in the fiber exchange with the sodium ions in the body fluid, and as a result, part of the fiber becomes sodium alginate. Since sodium alginate is water soluble, this ion exchange leads to the swelling of the fiber and the in-situ formation of hydrogel on the wound surface.

Figure 5.9 shows the photomicrographs of alginate fibers when placed in contact with water and saline, respectively. A highly swollen structure can be obtained by placing calcium alginate fibers in contact with aqueous solutions containing sodium ions.

Table 5.9 shows the gelling abilities of three types of alginate fibers with different M/G contents. The high G alginate fiber has about 70% G and 30% M, while

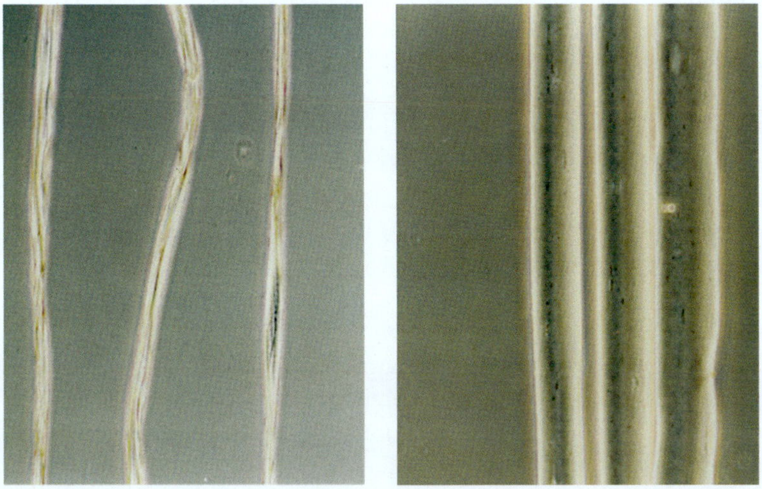

(a) (b)

Figure 5.9 Photomicrographs of alginate fibers, (a) wet in water; (b) wet in saline. Source: Adapted with permission from figure 7.2 in Qin [13].

Table 5.9 The calcium release, absorption capacities, and gel strength of three types of alginate fibers.

	Type of alginate fiber		
Test parameter	**High G fiber**	**Mid G fiber**	**High M fiber**
Ratio of M/G	About 0.4	About 1.6	About 1.8
Fiber calcium content	98.3%	96.9%	96.2%
Ca(II) content in contact solution (ppm)	317.5	450	560
Calcium release (%)	0.9%	1.43%	1.87%
Gel swelling ratio in water (g/g)	2.69 ± 0.27	6.0 ± 0.87	5.69 ± 0.39
Gel swelling ratio in 0.9% saline (g/g)	8.49 ± 0.62	14.51 ± 0.74	15.89 ± 0.65
Gel strength	85.2 g	25.8 g	32.4 g

the high M alginate has about 65% M and 35% G. After the ion-exchange process in solution A, the solution of the high G alginate fiber contains 317.5 ppm calcium ions, while the solution of the high M alginate fiber contains 560 ppm calcium ions, almost twice as much as the figure for the high G alginate fiber. These results clearly demonstrate that the high M alginate fiber exchanges ions more readily with the sodium-containing solution, and its gelling ability is much better than the high G alginate fiber [7].

The product design of high M Sorbsan dressing utilized the good gelling ability of the high M alginate fibers. Sorbsan wound dressing is constructed as a loose unneedled felt, after being used on wounds, the dressing can be easily removed from the wound bed with a rinse of warm saline solution, thus reducing the pain associated with the removal of wound dressings. On the other hand, the Algosteril dressing is made of high G alginate. Its ability to ion exchange with the wound exudate is limited and it is more difficult to form gel than Sorbsan. However, the Algosteril fibers are stronger than Sorbsan fibers when wet, and since it is in a needled format, they can be removed from the wound surface in one piece by using a forcep. During clinical use, both dressings have their own unique characteristics.

Tri-sodium citrate has a strong chelating capacity for metal ions, and in order to convert the calcium alginate fibers into a hydrogel, a 1.5% aqueous tri-sodium citrate solution was used to treat the high G and high M calcium alginate fibers, resulting in the formation of their respective hydrogels. From Table 5.9, it can be seen that since the high G alginate has a strong calcium ion binding ability, the gels formed from the high G alginate fiber have much higher gel strength than that of the high M alginate fiber. This high gel strength of the high G alginate also has an additional benefit during the production process. After extruding into calcium-containing coagulation bath, the high G alginate is able to form thread with high wet strength, hence it is easy to process than the high M alginate [7].

Table 5.10 shows the gel swelling ratio and the absorption capacities of Sorbsan, Algosteril· and Kaltostat. It is clear that Algosteril has worse performances than Sorbsan and Kaltostat. This is because Algosteril is a calcium alginate fiber wound dressing made of high G alginate. Compared to the high M Sorbsan, the high G alginate in Algosteril binds calcium ions firmly, and it is difficult for ion exchange and gelling to take place. Algosteril has poor gel-forming properties. The gel swelling ratio in 0.9% saline is only 5.2 g/g compared to 13.9 g/g for Sorbsan. Kaltostat is also

Table 5.10 The absorption behavior of Sorbsan, Algosteril, and Kaltostat.

	Type of alginate wound dressing		
Test parameter	**Sorbsan**	**Algosteril**	**Kaltostat**
Fiber calcium content	95%	98%	80%
Absorption capacities in solution A (g/g)	16.7	14.2	17.3
Gel swelling ratio in water (g/g)	2.2	1.8	7.7
Gel swelling ratio in 0.9% saline (g/g)	13.9	5.2	5.9

made of high G alginate, however, during the production process, part of the calcium ions in the Kaltostat fiber is already replaced by sodium ions, with the proportion of carboxylic acid group in calcium or sodium salt at 80/20. The 20% sodium introduced during the production process significantly increases the gel swelling ratio in water, and the gel swelling ratio for Kaltostat is roughly 4 times that of Algosteril. After the introduction of 20% sodium, the absorption capacities were also significantly increased, with Kaltostat absorbing 17.3 g/g solution A, as compared to 16.7 g/g for Sorbsan and 14.2 g/g for Algosteril.

Sodium ion content of the fibers is also important to the absorption and gelling properties of alginate fibers. Table 5.11 shows the absorption capacities and calcium release of three types of alginate fibers containing different levels of sodium ions [7]. It can be seen that as the content of sodium ions increases, there is a corresponding increase in the absorption capacities of the dressings. The gel swelling ratio in 0.9% saline increased from 8.49 g/g for the high calcium fiber to 18.58 g/g for the high sodium fiber. This result suggests that one way of improving the absorption and gelling properties of the alginate wound dressing is to introduce sodium ions into the fiber [9].

Table 5.12 shows the changes in fiber length after four types of alginate fibers were placed in distilled water and 0.9% saline solution [7]. It is clear that since the high G alginate has strong calcium binding and poor gelling ability, the fiber has a firm

Table 5.11 The absorption capacities and calcium release of three types of alginate fibers containing different levels of sodium ions.

	Type of alginate fiber		
Test parameter	**High calcium fiber**	**Mid calcium fiber**	**High sodium fiber**
Fiber calcium content	About 98.3%	About 76.5%	About 54.1%
Ca^{2+} content in contact solution (ppm)	317.5	230	136.2
Gel swelling ratio in water (g/g)	2.69 ± 0.27	24.05 ± 0.85	20.60 ± 2.66
Gel swelling ratio in 0.9% saline (g/g)	8.49 ± 0.62	13.42 ± 1.03	18.58 ± 0.94
Gel strength (g)	85.2	19.8	7.4

Table 5.12 Change of alginate fiber length in aqueous medium.

	Change of fiber length	
Type of alginate fiber	**Water**	**0.9% Saline**
High G high calcium alginate fiber	+0.99%	−6.14%
High M high calcium alginate fiber	−3.28%	−8.96%
High G high sodium alginate fiber	−9.22%	−12.00%
High M high sodium alginate fiber	−13.88%	−16.21%

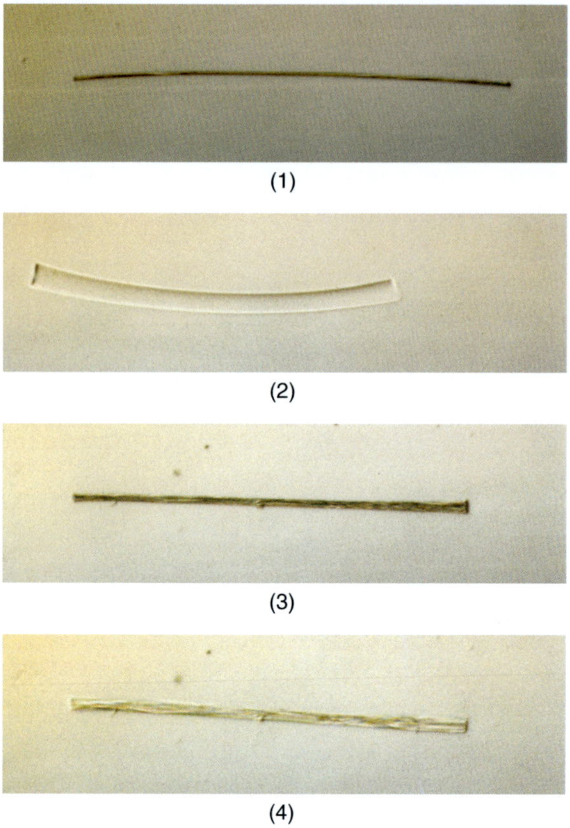

Figure 5.10 Change in fiber length before and after wetting with 0.9% saline, (1) high M calcium alginate fiber, dry; (2) high M calcium alginate fiber, wet; (3) high G calcium alginate fiber, dry; (4) high G calcium alginate fiber, wet.

structure and has a good dimensional stability when wet. On the other hand, the high M-type alginate fiber gels more easily, and as the fiber absorbs water during gelling, the oriented structure is disrupted, resulting in the contraction of the fiber. For similar reason, the fibers with high sodium content gel more easily and have significant shrinkage in fiber length when wet. Figure 5.10 shows the change in fiber length before and after wetting with 0.9% saline.

5.5 Summary

Alginate fibers have novel ion-exchange properties. As a polymeric acid, alginate can form salt with calcium, sodium, zinc, and many other metal ions. The calcium ions in the fibers, which are necessary for the formation of fibers in the production process, can be replaced by sodium ions when placed in contact with body fluid. The calcium alginates can be easily converted into mixed calcium/sodium or calcium/zinc alginate fibers by treating the fiber with aqueous solutions of Na_2SO_4 or $ZnCl_2$. The calcium/sodium alginate fibers are notably more absorbent than the calcium alginate fiber, while the calcium/zinc alginate fibers have the ability to deliver zinc ions when placed in contact with body fluid. The gelling ability is affected by the guluronic acid and mannuronic acid contents of the fiber, and the calcium and sodium contents of the fiber. High M alginate fibers exchange ions more readily than high G fibers, hence they have better gelling abilities than high G fibers. By introducing sodium ions into the fiber, the gelling ability and absorption capacities can be improved for the high G alginate fibers.

References

1 Chamberlain, N.H., Johnson, A., and Speakman, J.B. (1945). Some properties of alginate rayons. *J Soc Dye Colour* 61 (1): 13–20.

2 Dudgeon, M.J., Thomas, R.S., and Woodward, F.N. (1954). The preparation and properties of some inorganic alginate fibers. *J Soc Dye Colour* 70 (6): 230–237.

3 Haug, A., Myklestad, S., Larsen, B. et al. (1967). Correlation between chemical structure and physical properties of alginates. *Acta Chem Scand* 21: 768–778.

4 Johnson, A. and Speakman, J.B. (1946). Some uses of calcium alginate rayon. *J Soc Dye Colour* 62 (4): 97–100.

5 Mikołajczyk, T. and Wołowska-Czapnik, D. (2005). Multifunctional alginate fibers with anti-bacterial properties. *Fibres Text East Eur* 13 (3): 35–40.

6 Mo, L., Chen, J., Song, J. et al. (2009). Absorption of copper ions by alginate fiber. *Synth Fibers China* 38 (2): 34–36.

7 Qin, Y. (2004). Gel swelling properties of alginate fibers. *J Appl Polym Sci* 91 (3): 1641–1645.

8 Qin, Y. (2005). The ion exchange properties of alginate fibers. *Text Res J* 75 (2): 165–168.

9 Qin, Y. (2005). Calcium sodium alginate fibers. *Chem Fibers Int* 2: 98–99.

10 Qin, Y. (2006). The characterization of alginate wound dressings with different fiber and textile structures. *J Appl Polym Sci* 100 (3): 2516–2520.

11 Qin, Y. (2008). Alginate fibers: an overview of the production processes and applications in wound management. *Polym Int* 57 (2): 171–180.

12 Qin, Y. (2008). The gel swelling properties of alginate fibers and their application in wound management. *Polym Adv Technol* 19 (1): 6–14.

13 Qin, Y. (2016). *Medical Textile Materials*. Cambridge: Woodhead Publishing.

14 Qin, Y. and Chen, J. (2011). Absorption and release of zinc ions by alginate fibers. *J Text Res* 32 (1): 16–19.

15 Qin, Y. and Gilding, D.K. (1996). Alginate fibres and alginate wound dressings. *Med Device Technol* 11: 32–35.

16 Qin, Y., Cai, L., and Zhu, C. (2011). Antimicrobial properties of zinc alginate fibers. *J Text Res* 32 (2): 18–20.

17 Smidsrod, O. and Haug, A. (1972). Dependence upon the gel-sol state of the ion-exchange properties of alginates. *Acta Chem Scand* 26: 2063–2074.

18 Smidsrod, O., Haug, A., and Whittington, S.G. (1972). The molecular basis for some physical properties of polyuronides. *Acta Chem Scand* 26: 2563–2564.

19 Speakman, J.B. and Chamberlain, N.H. (1944). The production of rayon from alginic acid. *J Soc Dye Colour* 60: 264–272.

6

Applications of Alginate Fibers as Smart Woundcare Materials

6.1 Introduction

Wounds are defined as skin defects caused by mechanical, thermal, electrical, and chemical injuries, or by the presence of an underlying medical or physiological disorder. Wound dressings are materials used to cover the wounds. Many types of wounds occur in everyday life, such as mechanical injuries including abrasions, lacerations, acute bullet or knife cuts, bites and surgical wounds, and various types of burns caused by thermal, chemical, electrical, and radiational injuries. Other types of wounds such as chronic ulcerative wounds including pressure sores and leg ulcers occur more commonly among elderly people.

Wound dressings have been in use for as long as that of the conventional textile materials. The primary function of the wound dressings is to avoid strikethrough and to protect the wounded site from contamination and further injuries. Wound dressings may not help accelerate the healing process but they must not in any way delay the wound repair process through a number of mechanisms, such as adherence to wound bed, leaching toxic component, and causing wound maceration. Wound dressings need to be easy to apply and easy to remove. These traditional requirements for the wound dressings have largely been met by the various types of gauzes and various derivatives of gauze dressings, such as bandages, wax-impregnated gauze, woven, nonwoven, and knitted gauze. Nonwoven materials laminated with perforated plastic film have been used to reduce dressing adherence and to reduce the amount of fibrous residue, thus overcoming one of the main shortcomings of the traditional wound dressings [13].

In the 1960s, the science of wound dressings took a breakthrough, when Winter [32] reported the results of a study on the treatment of pig wounds in an occlusive condition. It was found that when the wound is kept in a moist condition as would have resulted from the application of an occlusive dressing, epithelialization of the wound surface occurred much faster than if the wound is kept in an otherwise dry condition, which was then regarded as the desirable condition. Further studies on human wounds confirmed that wound healing took place much faster when the wound is kept in a moist condition [12, 30]. These early studies provided the scientific and medical foundation for the modern wound management materials, which are designed to provide a moist healing environment for the wound.

Alginate Fibers and Wound Dressings: Seaweed Derived Natural Therapy, First Edition. Yimin Qin.

The so-called moist healing dressings were developed in order to create this moist interface between the wound and the dressing [24].

It is interesting to note that much of the research that led to the development of modern wound management products was concentrated on the "moist healing" principle, and in the 1980s and 1990s, many "moist healing" products, such as hydrocolloids, alginates, polyurethane foams, and hydrogel dressings, were developed and launched into the European and North American health care market, where the advent of the aging population together with the increased need for managing chronic ulcerative wounds provided a growing market for companies supplying wound care products, at the same time stimulating the research and development efforts for smart wound care materials that are more effective and more functional than traditional materials. Figure 6.1 shows the number of new wound care products listed in the British Drug Tariff during the 1990s. There was a clear surge in the number of new products from 1996 and onwards, when many new types of alginates, hydrogels, and other smart wound care materials became commercialized.

There have been many innovative approaches to further improve the clinical efficacy of the wound management materials and the effectiveness of the wound care products so as to increase the efficiency of the treatment process, especially in the case of chronic wounds. For example, many high-tech wound dressings aim to provide the best performance in terms of providing a large capacity for handling wound exudate, since the duration of a dressing can be extended by increasing the fluid handling capacity, and major savings in the cost of wound management can be achieved through the reduction of the number of changeovers, which takes up nursing time

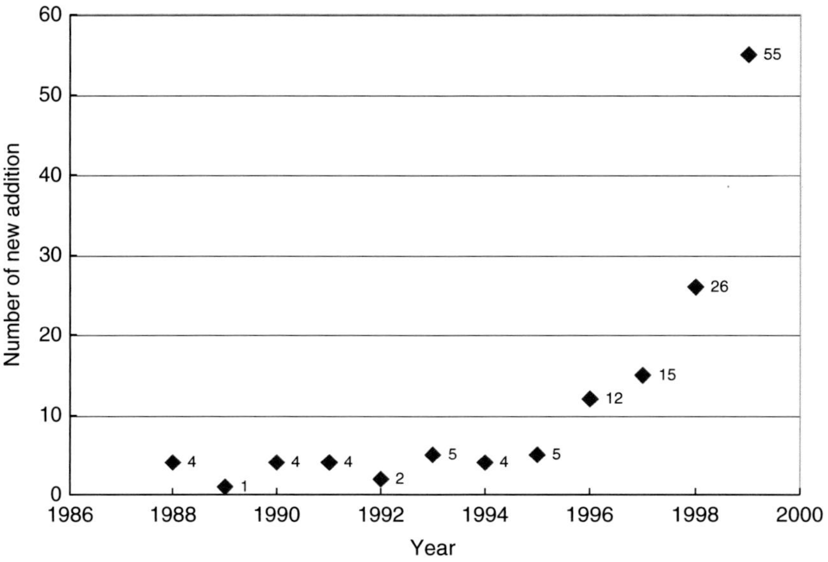

Figure 6.1 Number of new wound care products listed on the British Drug Tariff during the 1990s.

and the associated cost of sanitary materials and equipment. Wound dressings that can provide an interactive role to stimulate the rate of healing have also been developed, as a result of the advancement in the understanding of the mechanism of wound healing and the establishment of the optimum treatment methods for various types of wounds [28–30].

6.2 Functional Requirements of the Wound Dressings

As been mentioned previously, many types of wounds can be developed as a result of physical, chemical, mechanical, and biological damages to the skin and the underlying body. Wounds differ from one type to another, and therefore, the functional requirements are different for different types of dressings. For the majority of the dressings, the primary function is to avoid strikethrough and to provide a physical protection of the wound. Many wounds produce large quantities of exudate and the dressing must be able to absorb the exudate. As a basic requirement, the dressing should also be free of chemical and biological contamination and ideally should provide a protection against bacteria contamination of the wound. The basic functional requirements of the dressings have been known as follows [28]:

- Remove excess exudate from the wound surface
- Maintain a high humidity level at the wound/dressing interface
- Allow gaseous exchange
- Provide thermal insulation
- Be impermeable to microorganism
- Not shed fibers, leach out toxic substance nor provide a sensitivity or allergic reaction
- Allow removal without causing trauma to the tissue

In defining the requirements for wound dressings, it is also important to observe that for a specific type of wound, the requirement for the dressing is different at different stages of the healing process. According to their physiological conditions, wounds can be classified into five types, each with a symbolic color code, i.e. necrotic wounds (black), sloughy wounds (yellow), granulating wounds (red), epithelializing wounds (pink), and infected wounds (green). These five types of wounds differ in their physical appearance, the level of exudate, and the level of microbial contamination. Figure 6.2 shows a schematic illustration of the five types of wounds.

For the necrotic wounds, one of the main aims of applying a dressing is to facilitate the separation of the dead tissue from the underlining healthy tissue so as to enable a normal wound healing process. If exposed to a relatively dry atmosphere, as would be found in a hospital ward or in a centrally heated room, dead tissue rapidly loses moisture and becomes dehydrated. As it does so, it shrinks and progressively becomes olive-green or black in color. It is also hard and dry to touch. In this condition, autolysis is inhibited and separation of the necrotic tissue may be delayed indefinitely. For the treatment of necrotic wounds, it is essential that the dressings must be able to prevent the dehydration process. Indeed, they should be able to

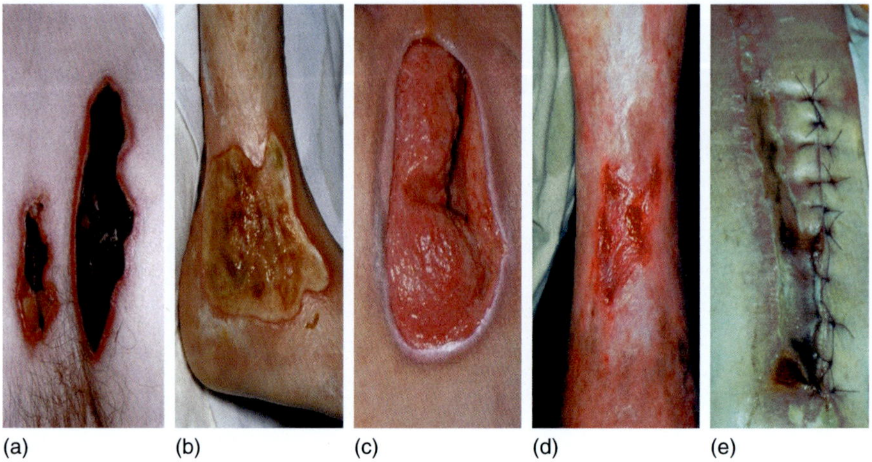

Figure 6.2 A schematic illustration of the five types of wounds, (a) necrotic; (b) sloughy; (c) granulating; (d) epithelializing; (e) infected. Source: Adapted from figure 7.1 in Qin [24].

facilitate a hydration process for the dead tissue so that the autolytic debridement process, in which the dead tissue is separated from healthy tissue, can take place. Necrotic wounds are ideally treated with hydrogels and/or hydrocolloid dressings.

For the yellow sloughy wounds, such as burns, leg ulcers, and pressure sores, where a necrotic covering has been removed, a glutinous yellow covering normally develops on the wound surface. This is not dead tissue, but a complex mixture of fibrin, protein, serous exudate, leucocytes, and bacteria. To treat this type of wounds, the sloughy mess must first be properly cleaned or debrided. If the slough is moist, an alginate dressing can be used to absorb the exudate. On the other hand, if the slough is dry, hydrocolloid or hydrogel dressings may be used.

Normal wound healing progresses from necrotic and sloughy phases to granulating phase, when the wound is covered by granulation tissue composed of collagen and proteoglycan in a complex of protein and polysaccharides, with salts and other colloidal materials. These produce a gel-like matrix that is contained within the fibrous collagen network with a highly vascular nature that gives it a characteristic deep pink color. Granulating wounds vary considerably in size, shape, and the amount of exudate that they produce. As a result, no single dressing will be suitable for use in all situations. Depending on the particular shape and the amount of exudate, alginate, hydrocolloids, and hydrogels may be used.

The final phase of the wound healing process is the epithelialization of the wound surface. With a few exceptions, superficial or epithelializing wounds tend not to produce large quantities of exudate. Traditionally, these wounds have been dressed with paraffin gauze and cotton tissue but sometimes alginate and hydrocolloids can be used. It is important to recognize that at this stage, the tissue is soft and fragile and therefore, any dressings used on these wounds must not adversely disturb the delicate tissue structure. In particular, dressings must not unnecessarily adhere to the wound. For epithelializing wounds, alginate, hydrocolloid, vapor-permeable film, silicon-coated film, and knitted viscose gauze may be used as the primary wound dressings.

Infected wounds tend to generate a large amount of exudate and also unpleasant odor. The treatment of infected wounds therefore comprises absorbing exudate, containing odor, and controlling microbial contamination. Dressings that in part contain activated charcoal can be used to absorb odor, while the odor-generating bacteria can be controlled by using dressings with an antimicrobial function, such as those with chlorhexidine, silver compounds, or iodine.

The main functions of wound dressings are summarized below [21, 24].

1. Fluid control
   The ability to absorb fluid from a highly exuding wound or to donate moist to a dry wound is one of the main functions of a wound dressing.
2. Odor management
   Wound quite often produces unpleasant and obnoxious odor. When this occurs, dressings must be able to contain the odor.
3. Microbial control
   For infected wounds, it is important that bacteria are contained by appropriate methods.
4. Physical barrier
   One of the principal functions of a wound dressing is to avoid strikethrough. In addition to its aesthetic purposes of hiding the wound, wound dressings also help to separate the wound surface from the atmosphere and prevent the wound from bacteria contamination and further physical damage to the tissue.
5. Space filler
   For deep cavity wounds, it is important that the wound is kept open by filler materials, so that the healing process can take place from bottom upwards, and unnecessary closure of the wound before the whole cavity is healed can be prevented.
6. Debridement
   The removal of dead necrotic tissue is essential for facilitating the normal wound-healing process. Wound dressings can accelerate the debridement process by providing the appropriate moisture, pH, temperature, and other conditions that are ideal for the autolytic debridement process.
7. Hemostatic effect
   For acute surgical wounds and traumatic wounds, it is important that bleeding is stopped as early as possible to prevent blood loss. Appropriate wound dressings can help blood clotting.
8. Low adherence
   Adherence of the whole or part of the dressing to the wound surface is a major problem in wound management, often causing trauma on removal of the dressing. A low-adherent dressing can help lower or eliminate adherence to wound.
9. Scar reduction
   For large wounds, scar formation presents a major aesthetic problem for the patient. Any dressing that can reduce or prevent scar formation can give a great benefit to the patient.

10. Metal ion metabolism
    A number of metal ions, such as iron, zinc, copper, magnesium, and selenium, play important roles in cellular activities. Deficiency in any metal ions delays wound healing. Apart from systemic intake of these metal ions, they can be administered through the topical use of appropriate wound dressings.
11. Pressure regulation
    The arterial and venous blood flow requires an appropriate pressure, and pressure sore develops when body parts are subjected to a sustained high pressure. Elastic bandages can help regulate body pressure and assist the wound-healing process.
12. Wound-healing acceleration
    Wound healing is a complex physiological process. Wound dressing often plays a minor role in the overall rate of the wound healing process. However, when combined with a number of factors, the appropriate use of dressings can accelerate the wound-healing process.

6.3 Modern Advanced Wound Dressings

With the improvement in the understanding of the healing process and mechanism, and with the development in modern material technologies, there have been many diversified approaches to the development of modern woundcare materials. Increasingly, the traditional textile gauzes have been replaced by various types of modern dressings in such diverse forms as powders, solutions, gels, films, membranes, foams, fibers, ribbons, ropes, felts, and a combination of the various types of dressings into the so-called wound dressing systems. Many types of materials have also been widely used in making the dressings, typically polysaccharides such as alginate, carboxymethyl cellulose, pectin, chitin, chitosan, and their derivatives, and polyurethane elastomers in the form of film, foam, and membrane.

It is often difficult to divide a line between the so-called traditional dressings and the modern high-tech dressings. As mentioned before, the evolution of the modern dressings began with the discovery of the "moist healing" concept, and many high-tech wound dressings are often able to control the level of moist at the interface between the wound and the dressing so as to create a "moist but not wet" condition. Since the 1990s, more and more advanced wound dressings have been developed by a large number of companies in the wound management industry and launched into the global healthcare market. As an illustration of the diversity of modern smart woundcare materials, the British Drug Tariff has 10 categories of advanced wound dressings, as can be seen below.

1. Low adherence dressings
2. Gauze impregnated dressings
3. Dextranomer paste pad and dressing
4. Alginate dressings
5. Hydrocolloid dressings
6. Hydrogels

7. Vapor permeable adhesive film dressings
8. Polyurethane foam dressings
9. Zinc paste bandages
10. Iodine containing dressings

These products are often used alone or in combination on wounds with many diversified physiological backgrounds and at different stages of the healing process. In order to satisfy the requirement of the healing process, many types of smart materials and technologies have also been developed. Some of these are introduced below.

6.3.1 Chitin and Chitosan Fibers and Wound Dressings

Chitin, poly-1,4-2-acetamido-2-deoxy-β-D-glucose, is the second most abundant natural polymer existing widely in cell walls of fungi and crustacean shells. Chitin is commercially produced from the shell waste of crabs, shrimps, and krills through a series of deproteinization and demineralization to remove the protein and minerals, which together with chitin, form the composite structure of the shells. The dry mass of shell waste typically contains about 15–25% of chitin (MUZZARELLI R A A. Chitin. New York: Pergamon Press, 1977).

Chitin has long been known as being able to accelerate the wound healing process. It has been shown that by applying chitin dressings, the wound-healing process can be accelerated by upto 75%. Chitin fiber was first reported in 1926 [14, 15], when German scientists succeeded in making an artificial silk that resembles the texture of natural silk. Chitin is however by its chemical and physical nature very difficult to dissolve. Although a large number of solvents, such as concentrated mineral acids, trichloroacetic acid, and formic acid have been used to dissolve chitin, the dissolution process is often complicated and impractical for large-scale fiber production [1].

In the 1980s, following extensive development on the solvent system, a new solvent for chitin was developed which offers the opportunity for the easy processing of chitin into fibers. By treating chitin first with *p*-toluene sulfonic acid in i-propanol, chitin can be easily dissolved in dimethyl acetamide (DMAc) containing a small amount of lithium chloride. The chitin solution in DMAc-LiCl can be extruded into a coagulation bath of either water or methanol solution to form fibers.

Chitosan is the deacetylated derivative of chitin. Like chitin, chitosan is also known to have wound-healing acceleration properties and a number of studies have shown that chitosan fibers have unique properties as a suture and wound dressing material. Chitosan can be easily dissolved in aqueous solutions of almost all the organic and inorganic acids because of the primary amine group on the C-2 position of the glucose residue. Chitosan fibers can be made by first dissolving it into an aqueous acidic solution and then extruding the solution through fine holes into a coagulation bath of a dilute alkali solution. Chitosan precipitates out in the form of a filament which can be washed, stretched, and dried to form fibers for the production of wound dressings [7].

Since chitosan is chemically the deacetylated form of chitin, by acetylating chitosan with acetic anhydride, it has been found that the chitosan fibers can be converted to chitin fibers, which have similar properties to that of the natural chitin fiber [23]. Both chitin and chitosan fibers have good mechanical properties, with fiber tenacity in the region of 1.5–2.5 g/decitex and elongation at break at 8–20%. These fibers are highly hydrophilic and are biocompatible, biodegradable, and nontoxic, providing unique raw materials for the production of advanced wound dressings.

6.3.2 Superabsorbent Cellulosic Fibers

In order to produce highly absorbent fibers, cellulosic fibers such as cotton and viscose rayon can be treated with chloro-acetic acid to make carboxymethyl cellulose fibers. As can be seen in Figure 6.3, when the cellulosic fibers are partially carboxymethylated, the carboxylic groups in the fibers are capable of absorbing a large amount of water, and the fibers are capable of a high degree of swelling when wet in water. By controlling the degree of substitution, it is also possible to maintain their fibrous form when in contact with water [20, 22].

The carboxymethylation process can be applied to the solvent-spun Tencel fibers, in which case, the nonwoven dressing can retain the soft and fine features of the solvent-spun fibers and also at the same time possess the superabsorbency that is derived from the carboxymethylation treatment. This type of product has the ability to absorb fluid directly into the body of the fiber, thus significantly increasing the volume of fluid that can be absorbed. In clinical circumstances, the removal of a large volume of exudates may lead to a decrease in the number of microorganisms on the wound surface. Also, in the presence of wound exudates, the carboxymethylated cellulosic dressings can form a cohesive gel sheet, which facilitates their use under compression bandages. The non-adherent features of these dressings lead to significantly less pain on dressing change.

As can be seen in Figure 6.4, upon contact with wound exudates, the carboxymethylated cellulosic fibers take the liquid up into the structure of the fibers themselves, instead of holding it within a web of gelled fibers. This property results in a range of clinical benefits such as:

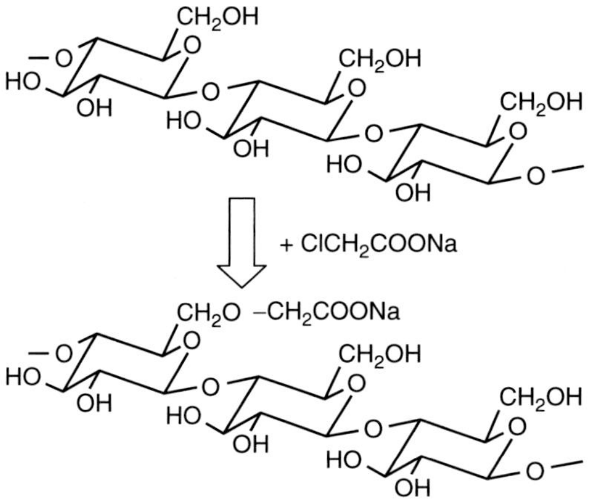

Figure 6.3 Chemical structure of cellulose and partially carboxymethylated cellulose.

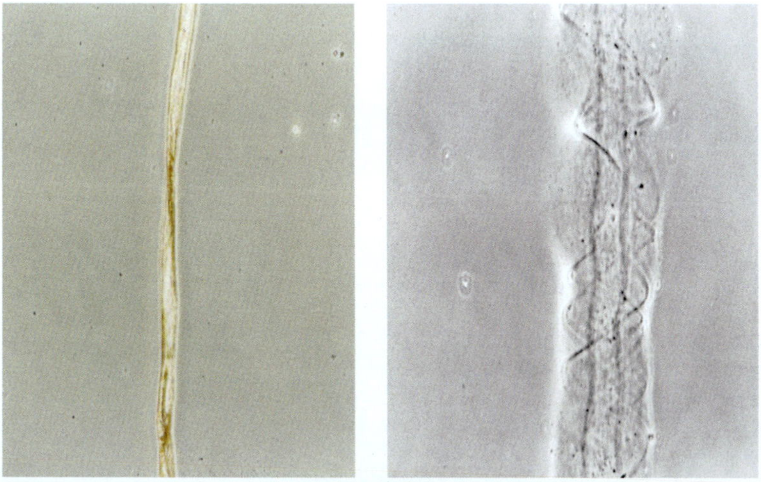

Figure 6.4 The dry (a) and wet (b) structure of partially carboxymethylated cellulosic fiber. Source: Adapted with permission from figure 5.2 in Qin [24].

- Superior exudate absorption and retention
- Improved handling characteristics
- Enhanced vertical wicking, which minimizes the potential for maceration

6.3.3 Polyurethane Film and Foam

Polyurethane-based semi-permeable polymeric coatings have long been used in the textile industry for making waterproof breathable clothing. As wound management material, a semi-permeable film allows gaseous exchange between the wound and the surrounding environment, while preventing air born bacteria from contaminating the wound. The semi-permeable nature of the film allows a high rate of moisture vapor transmission through the film, thereby reducing the build-up of wound exudate under the film. These films are highly comfortable and convenient to use and they can permit constant observation of the wound. They allow gaseous and moisture exchange but provide a barrier against water and microorganism.

Modern semi-permeable film dressings are usually made of polyurethane of various compositions and the moisture vapor transmission rate can be modified by modifying the polymer and film structure. Moisture vapor transmission rate of 3000 $g/m^2/24$ hours or greater can be achieved with the latest high tech films.

The semi-permeable film dressings can be used for surgical wounds and for protecting the site of insertion and in-dwelling catheter. They are also used in the treatment of superficial partial-thickness burns and early decubitus ulcers. These dressings are light and resilient, ideal for the prevention of skin damage against friction. They are also widely used as a secondary dressing in conjunction with hydrogels, hydrocolloids, and alginates. Figure 6.5 shows an illustration of a wound covered with a piece of semi-permeable polyurethane film.

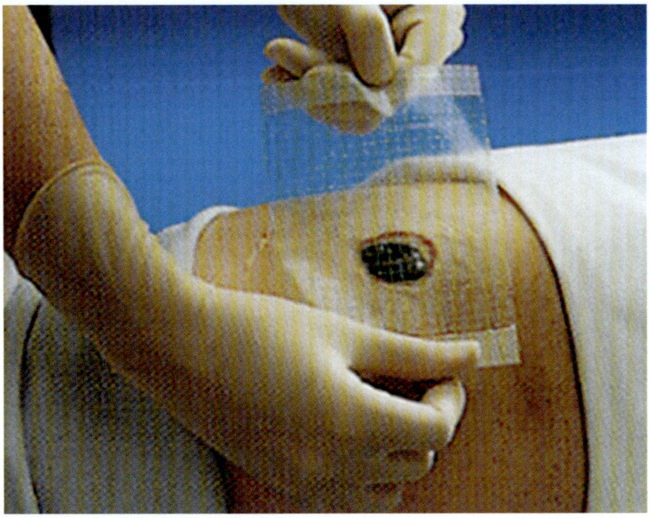

Figure 6.5 An illustration of a wound covered with a piece of semi-permeable polyurethane film.

Polyurethane foams are soft and porous materials that can be used for wound-contacting primary dressings or as secondary dressings, utilizing their relatively high strength and flexibility. The polyurethane can be made from hydrophobic or hydrophilic monomers, resulting in foams with varying characteristics of porosity and fluid handling capability. When placed on wet wounds, fluid is absorbed into the foam by capillary action and transferred across the dressing. When placed on relatively dry wound, the polyurethane backing layer reduces moisture vapor loss and helps to prevent dehydration of the wound surface.

6.3.4 Hydrogels

Hydrogels are cross-linked polymeric networks swollen in biological fluid. They are widely used in drug delivery and tissue/organ repairs. Two types of hydrogels are usually available in wound management products, i.e. sheet hydrogels and amorphous hydrogels. With the sheet hydrogel, the hydrophilic polymers, typically polyacrylamide or polyethylene oxide, are partially cross-linked to form a membrane with sufficient water holding hydrophilic sites. Typically, a hydrogel contains about 96% water. When applied to the wound sites, such as dermabrasion, minor burns, and skin donor sites, they relieve pain and reduce trauma both on the application and removal of the dressings. Hydrogels can be dried to form the dehydrated hydrogel, which has a higher absorption capacity than the hydrated film. These dressings are ideal carriers for antibiotics and placental growth factors. Figure 6.6 shows an illustration of a piece of sheet hydrogel.

Amorphous hydrogels can be made from a number of water-soluble polymers such as cross-linked carboxymethyl cellulose, modified starches, alginate, and pectin. Unlike the sheet hydrogel, which takes up a three-dimensional structure,

Figure 6.6 An illustration of a piece of sheet hydrogel.

these gels are thick viscous fluids. When absorbing wound exudate, they swell until they lose all the cohesive properties.

Amorphous hydrogels are excellent donors for water. They are ideally used on dry and sloughy wounds whereby the water-donating properties assist the autolysis process for the debriding and cleansing of slough and black necrotic tissue from the underlying healthy tissue. Amorphous hydrogels are also widely used in the treatment of cavity wounds, where their easy-flowing properties can be used for the packing of deep cavity to avoid wound closure from side to side.

6.3.5 Hydrocolloids

Hydrocolloid wound dressings are among the first type of high-tech modern dressings. These dressings are typically made of hydrophilic polymeric granules dispersed in an elastic adhesive matrix. Typically, the hydrophilic granules are hydrophilic polymers such as sodium carboxymethyl cellulose, pectin, gelatin, and sodium alginate, while the adhesive matrix is typically poly-isobutylene. The hydrocolloid and the adhesive form a homogeneous mixture such that the granules are uniformly dispersed in the adhesive matrix. The final dressing is usually composed of the hydrocolloid matrix cast on a sheet of polymeric film or membrane (occlusive or semi-permeable for different applications). On contact with wound exudate, the hydrophilic granules absorb the fluid to form a hydrogel, while the adhesive material provides a tack that keeps the dressing adhered to the wound. Since the moisture transportation is relatively slow through the hydrocolloid matrix, the wound surface is covered by a moist contacting layer which assists the healing of the wound. The dressing is easy to remove and it provides a good protection

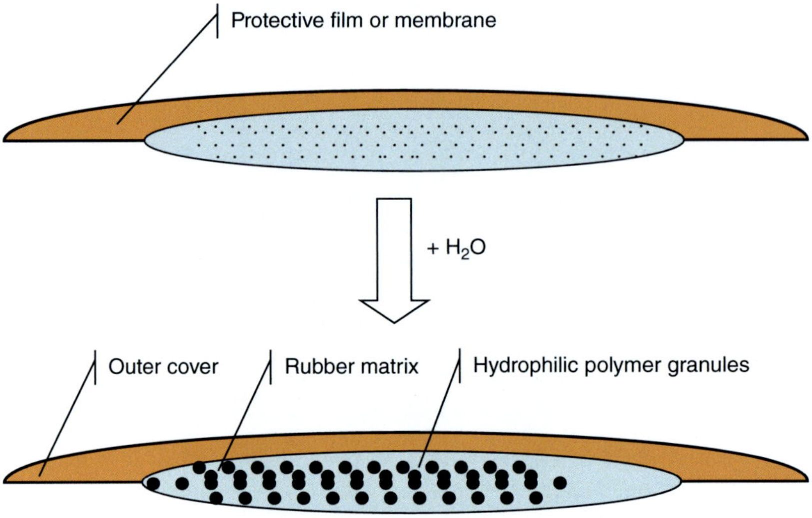

Figure 6.7 An illustration of the basic structure of hydrocolloid wound dressing.

against bacteria contamination of the wound. Figure 6.7 shows an illustration of the basic structure of hydrocolloid wound dressing.

6.3.6 Activated Carbon

The production of wound odor can represent a major problem for patients and their carers. Wounds most commonly associated with odor production include leg ulcers and fungating (cancerous) lesions of all types. The smell from these wounds is caused by a cocktail of volatile agents that includes short-chain organic acids, such as *n*-butyric, *n*-valeric, *n*-caproic, *n*-haptanoic and *n*-caprylic acids, produced by anaerobic bacteria, together with a mixture of amines and diamines such as cadaverine and putrescine that are produced by the metabolic processes of other proteolytic bacteria.

The most effective way of dealing with malodorous wounds is to prevent or eradicate the infection responsible for the odor. This may be achieved in a number of ways. The administration of systemic antibiotics or antimicrobial agents may be effective in some cases, but often the nature of the wound is such that it is not possible to achieve an effective concentration of the antibiotic at the site of infection by this method, particularly in the presence of slough or necrotic tissue. If the formation of the odor cannot be prevented, it may be necessary to use a dressing that can absorb the smell. Dressings containing activated carbon fibers can be used for the treatment of these malodorous wounds.

Activated charcoal cloth is produced by carbonizing a suitable cellulose fabric by heating it under carefully controlled conditions. During this process, the surface of the carbon breaks down to form small pores. These greatly increase the effective surface area of the fibers and hence their ability to remove unpleasant smells, since the molecules that are responsible for the production of odor are attracted to the surface of the carbon and are held there by electrical forces.

6.3.7 Low Adherent Dressings

One of the major problems with traditional gauzes is the difficulty in removing them upon healing of the wound. Fibrous matters from the dressing tend to mix up with exudate and blood. When the wound heals and dries, the fibers often get stuck with the dry eschar. Bleeding of the new tissue is common in these cases and the trauma in removing traditional cotton gauzes from large wounds often poses a serious problem.

Low adherent dressings were developed in order to prevent the above problem by placing a primary wound contacting layer that separates the exuding wound with the absorbent dressing layer. A typical low adherent dressing is the so-called tulle gras, invented during the first world war. These dressings are usually made of cotton gauzes impregnated with a coating of paraffin, which provides a low adherence to the wound surface. Quite often, the paraffin gauzes can be medicated to enhance their performances.

For heavily exuding wounds, the poor absorption properties of the tulle gras tend to reduce the fluid uptake and skin maceration occurs as a result of the build-up of the fluid under the dressing. Other forms of low adherent dressings were developed in order to improve this problem. Most of the modern low adherent dressings are perforated polymeric films, with the pores acting as the channel for the exudate to pass through to the absorbent layer. Knitted gauzes made of continuous viscose rayon filament as shown in Figure 6.8 are also used as low adherent dressings for the treatment of such heavy exuding wounds as leg ulcers. It is however usually a problem area for these dressings when the new tissue grows into the pores of the perforated film or knitted structure, causing tissue damage on removal of the

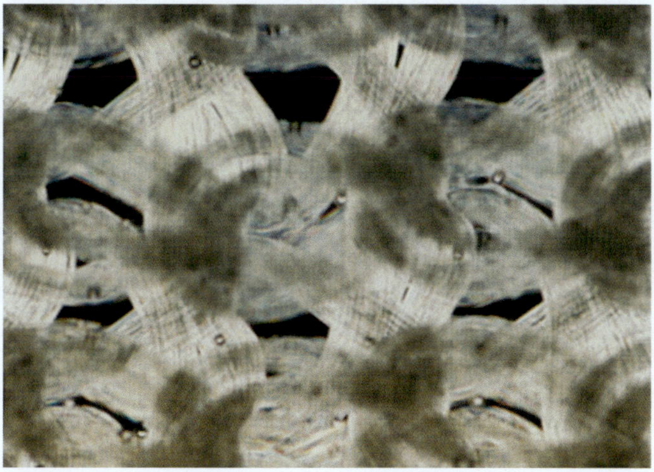

Figure 6.8 Photomicrograph of a low-adherent dressing made of knitted viscose rayon fabric.

dressing. In these respects, modern dressings such as hydrogels, hydrocolloids, and alginate offer far superior properties both as highly absorbent dressings and for the low adherent properties. Silicone-coated fabric can also be used to reduce adherence.

6.3.8 Composite Wound Care Products

After many years of research and development, especially since the widespread use of "moist healing" products in the 1990s, modern wound dressings are now much more functional and smart than the traditional products such as cotton gauzes and absorbent swabs. Compared to traditional products, the new generation of smart woundcare products tends to be easy to use and cost-effective. In achieving these functionalities, modern woundcare products also adopt many novel composite structures. For example, in the management of leg ulcers, the four-layer system has proven to be effective in applying compression as well as exudate management to chronic leg ulcers, which is achieved by first applying a wool bandage from the base of the toes to just below the knee joint, followed by the application of a crepe bandage and then an elastic compression bandage, and finally wrapped in a cohesive layer.

Many other types of composite woundcare products have also emerged. In general, these products are composed of three key components, i.e. the wound contact layer, the functional layer, and the retention layer. Figure 6.9 shows a schematic illustration of modern composite woundcare products.

In the three-layered modern composite wound dressings, the contact layer aims to provide a low or non-adherent interface between the wound surface and the dressing. This should allow wound exudate to pass into the functional layer while at the same time, being able to prevent the adherence of the dressing to a drying surface at the end of the healing process. The contact layer also stops the release of loose fibers or particles from the dressing into the wound site. As can be seen in Figure 6.10, polyamide nonwoven can be used as wound contact layer to prevent the direct contact between wound bed and the functional layer of activated carbon with silver. Other materials that have been used include perforated plastic films, knitted viscose filament yarn, silicone gel, etc.

The functional layers vary greatly for different products because of the variations in their intended uses. For the majority of wounds, the main issues in wound management are to contain wound exudate, control microbial growth, and combat

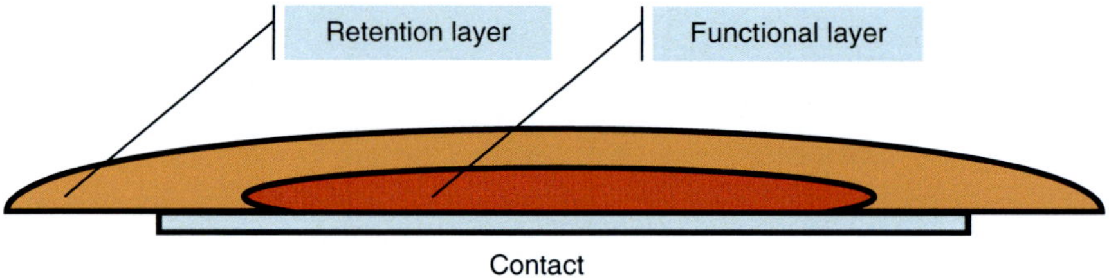

Figure 6.9 A schematic illustration of modern composite woundcare products. Source: Adapted with permission from figure 7.4 in Qin [24].

Figure 6.10 An example of contact and functional layers.

odor. In these cases, a layer of superabsorbent material, antimicrobial substance and activated carbon fabric, respectively, can be laminated into the wound dressing to provide the functional layer.

The retention layer is used for two main purposes, i.e. to secure the dressing onto the wound edge and to provide physical protection of the wound surface. Polyurethane films and hydroentangled nonwoven fabrics are often used as retention layers. These materials are soft and flexible, thus making it easy to apply the dressings onto curvy areas of the body, such as around the shoulders and elbow. In addition, these materials are breathable, thus allowing oxygen to penetrate into the wound, and allow moisture to evaporate from under the dressing, thereby extending the duration of the product. Figure 6.11 shows an example of absorbent layer secured by a retention layer made of hydroentangled nonwoven fabric.

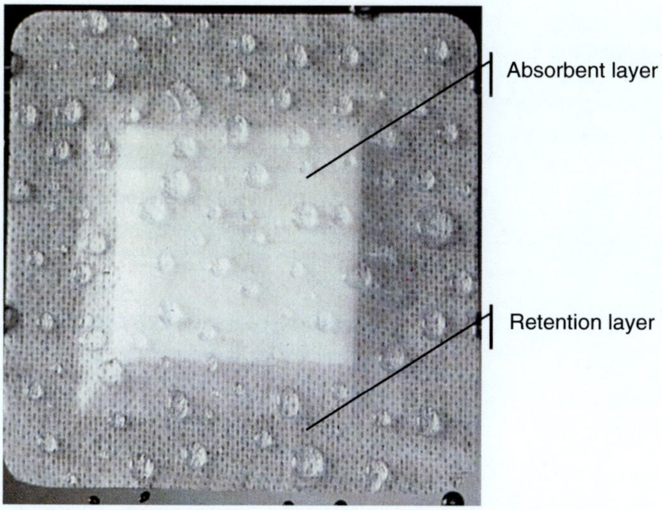

Figure 6.11 An example of functional and retention layers.

6.3.9 Antimicrobial Wound Dressings

The spread of antibiotic-resistant strains of microorganisms such as methicillin-resistant *Staphylococcus aureus* (MRSA) represents an ever-increasing threat to the health of vulnerable people throughout the world who are obliged to spend extended periods in healthcare facilities. The organism is also responsible for increasing the financial burden placed on such centers and the wider community at large, with the result that precious financial resources are diverted from other areas of need to deal with the consequences of infection. In order to overcome microbial infection, many dressings now exploit the "bioactive" properties to promote healing and control infection. These include the now well-known sustained release iodine and silver dressings. For example, Actisorb Plus is an activated charcoal cloth impregnated with silver. It is capable of absorbing bacteria, which are then inactivated by the silver ions in the dressing.

Another product, Acticoat, utilizes novel silver-coating technologies in a dressing designed to prevent wound adhesion, control bacterial growth and facilitate burn woundcare. Acticoat dressing consists of a rayon/polyester nonwoven core, laminated between layers of silver-coated high-density polyethylene mesh. This product can provide an effective antimicrobial barrier for up to 3–5 days against 150 pathogens, including both MRSA and vancomycin-resistant *Enterococci* (VRE).

As illustrated in Figure 6.12, novel antimicrobial wound dressings can also be made by blending calcium alginate fibers with silver-containing X-static fibers, which are nylon fibers coated with a layer of metallic silver on the fiber surfaces. In this system, calcium alginate fibers provide the high absorbency and gelling ability, while the silver-containing fibers provide the sustained release of silver ions and hence the antimicrobial properties of the product.

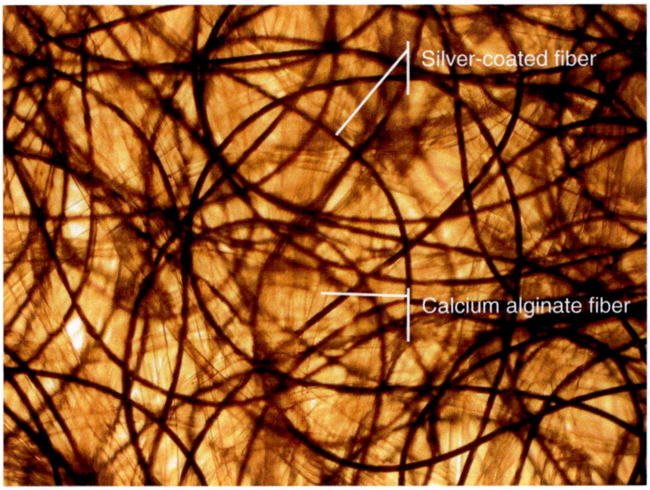

Figure 6.12 A composite dressing made of alginate and X-Static fibers. Adapted with permission from figure 11.2 in [24].

Because of the broad-spectrum antimicrobial efficacy and the relatively low toxicity of silver ions, there are now many attempts to combine silver ions into wound dressings, resulting in a large number of commercial silver-containing wound dressing products now available in the global wound management market [25].

6.3.10 Interactive Dressings

Interactive dressings are those products that can interact with cells or matrix proteins in the wound bed to promote healing. For example, alginate wound dressings are highly absorbable and biodegradable dressings derived from seaweed. As well as controlling exudate by ion exchange, alginate can also exert a bioactive effect by activating macrophages within the chronic wound bed to generate pro-inflammatory signals, such as tumor necrosis factor (TNF)-alpha, interleukin (IL)-1, -6 and -12. This may then initiate a resolving inflammatory response characteristic of healing wounds. *In vitro* studies have demonstrated that some dressings containing alginate can activate macrophages, as evidenced by their increased production of TNF-alpha.

Another example of interactive dressing is Promogran, which is a sterile, freeze-dried matrix made up of collagen and oxidized regenerated cellulose. In the presence of wound exudate, the matrix absorbs liquid and forms a soft, conformable, totally biodegradable gel, which re-balances the wound environment. The gel binds and inactivates matrix metalloproteinases, which when present in excessive levels, have a detrimental effect on wound healing as they damage regenerating tissue. The gel also binds growth factors secreted by macrophages and fibroblasts in the wound bed, protecting them from degradation by these proteases. As the gel is digested during the course of healing, the growth factors are released back into the wound bed in their active form, thereby promoting the healing process.

6.3.11 Tissue-Engineered "Skin Equivalents"

Surgical grafting of split-thickness autologous skin is the standard method for rapid closure of full-thickness burn wounds. Modern tissue engineering has now made possible grafts using either sheets of fibroblasts in a biodegradable matrix or cultured keratinocyte sheets. Superior results can be obtained when both dermal and epidermal components are combined in a bi-layer skin equivalent.

6.3.12 Cell-Containing Matrices

Artificial skins can be developed by combining cells with either synthetic matrices, such as polyglycolic acid mesh, or natural biological substrates such as collagen and glycosaminoglycans. For example, Dermagraft was developed by using a polyglactin-910 surgical mesh seeded with human dermal fibroblasts. It has been found that this product can allow re-vascularization and support human meshed split-thickness skin grafts.

6.4 Applications of Alginate Fibers in Functional Wound Dressings

Alginate fibers have a unique ion-exchange property. On contact with wound exudate, calcium ions in the fiber exchange with sodium ions in the body fluid and as a result, part of the fiber becomes sodium alginate. Since sodium alginate is water soluble, this ion exchange leads to the swelling of the fiber and the in-situ formation of gel on the wound surface. This unique property makes alginate fiber one of the ideal material for the production of "moist healing" wound dressings, and after much development during the 1980s and 1990s, many types of alginate fibers and wound dressings are now available, utilizing the diversified properties of the different types of alginate extracted from different sources of seaweeds and the availability of many types of salts of alginate, such as zinc and silver alginate, which are used for zinc-deficient people and for antimicrobial properties, respectively. Due to their unique properties and the fact that the dressings can be used in the dry form or hydrated form, alginate dressings can be used for a wide range of wounds, providing a cost-effective treatment that involves a minimum number of dressing changes. Figure 6.13 shows an illustration of the three types of alginate wound dressings

The properties of alginate fibers and wound dressings can be modified in many ways. For example, in order to make the alginate fibers more absorbent, sodium ions can be introduced into the calcium alginate fibers through chemical treatment. In this process, the calcium alginate fibers can be first washed with hydrochloric acid to replace part of the calcium ions with hydrogen ions, which are then replaced with sodium ions by a treatment with sodium carbonate or sodium hydroxide [26]. Highly absorbent calcium sodium alginate fibers can also be made by treating the fibers with aqueous solutions containing different amount of Na_2SO_4.

When the sodium content increases, either during the production process, or when placed on exuding wounds, alginate fibers are capable of holding a large amount of water within the fiber structure. This is important in two respects.

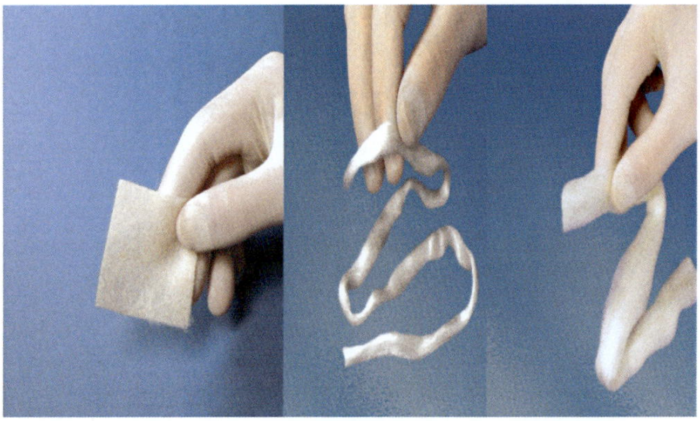

Figure 6.13 An illustration of the three types of alginate wound dressings.

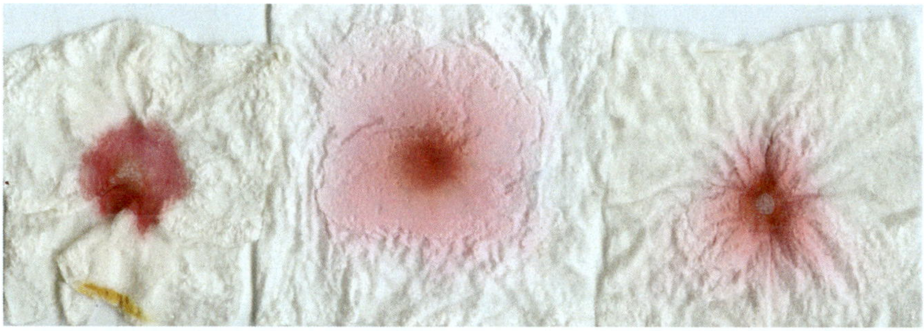

Figure 6.14 The spreading of liquid on nonwoven fabrics made of sodium alginate, alginic acid, and calcium alginate fibers.

First, as the fibers hold more water, the dressings are capable of absorbing more wound exudate, hence extending the duration of the dressing. Second, when the fiber absorbs water into the fiber structure and swells, the spaces between the fibers in the dressing are closed, thus prohibiting liquid from lateral spreading and preventing the maceration of the areas surrounding the wound surface. As can be seen in Figure 6.14, when 5 ml of normal saline is dropped onto the nonwoven structure, the sodium alginate fabric is capable of holding the liquid within a very narrow area, showing a much better "gel blocking" property than the calcium alginate and alginic acid fabric.

6.5 Development of Alginate Wound Dressings

As far as woundcare is concerned, seaweeds have long been used by sailors to cover wounds, and they are known as "mariner's cure." During the World Wars in Europe, dried sea moss dressings were sent to field hospitals to treat wounded soldiers [11, 16, 31]. The first person in modern times to recognize the potential value of alginates in surgery and wound management was George Blaine, a major in the Royal Army Medical Corps. He showed them to be absorbable in tissue, sterilizable by heat, and compatible with penicillin [2]. He also described how he had used alginate films clotted in situ for the treatment of wounds and burns in troop ship hospitals in the Far East and described the use of alginate, sometimes in combination with plasma as an alginate-plasma film, as "puncture patches" over scleral defects.

During a subsequent assessment of the use of alginate as hemostats and wound dressings, Blaine reported their apparent lack of toxicity following a series of animal studies in which fibers were implanted into animal tissues, and gels made from alginate were used to treat experimentally produced burns [3]. Clinical studies followed, and the successful use of alginate-derived materials in aural surgery and neuro surgery was reported by Passe and Blaine [19] and Oliver and Blaine [18], respectively.

Other more general applications were described in 1948, when the results of a 3-month trial into the use of alginate in the casualty department of Croydon Hospital were reported by Bray et al. [6]. In this study, alginates in the form of films, wool, gauzes, and clots (formed in situ by mixing sterile solutions of calcium chloride and sodium alginate) were applied to a wide range of wounds, including burns, lacerations, ulcers, and amputations. In all cases, healing was rapid and uneventful.

According to the results of a survey carried out by Stansfield and reported by Blaine [4], in the late 1940s and early 1950s, alginates were being used in some 70 hospitals over a range of surgical specialties. Overall, they were found to be highly satisfactory in use. Where criticisms were recorded, they were directed mainly at the poor absorption properties of the material and its consequent tendency to induce fistula formation. It was noted that most of these criticisms related to cases in which the product had been used as packing for large cavities or dead spaces, a function for which it was never originally intended.

Following the early work of Blaine and others, a number of commercial medical alginate products were produced, including an absorbable swab called Calgitex. When the large-scale manufacture of alginate fiber ceased in the early 1970s, due probably to the emergence of synthetic fibers, this product was discontinued, owing to the high cost of production.

The advancement of "moist healing" practice and related products revived the use of alginate in wound management, especially for the care of chronic ulcerative wounds such as pressure sores and leg ulcers. In the early 1980s, the first modern alginate wound dressing was launched and the first clinical report recording the use of Sorbsan™ was published in 1983 when Fraser and Gilchrist [8] and Gilchrist and Martin [9] described their experiences with the dressing in the management of foot disorders and a variety of skin lesions, following a clinical evaluation in a group of hospitals in the Sunderland area in the United Kingdom. The results of these studies were very positive and supported the findings of Blaine some 40 years earlier. Further papers described the use of Sorbsan in the management of problem wounds including infected traumatic wounds and leg ulcers [17, 27].

In 1986, the second alginate wound management product, Kaltostat, was launched and in 1988, alginate wound dressings gained widespread clinical acceptance when Sorbsan was included in the Drug Tariff in the United Kingdom. Sorbsan is also approved by the FDA (D.C. Number K881854).

Many clinical applications in the 1980s and 1990s helped to establish the many benefits of alginate wound dressings. For example, Groves and Lawrence [10] found that alginate dressing was able to significantly reduce blood loss from skin graft donor sites. When in contact with blood or exudate, alginate wound dressing releases calcium ions in exchange for sodium, thereby increasing the local calcium ion concentration and stimulating both platelet activation and whole-blood coagulation. The nonwoven construction of these dressings acts as a matrix for clot formation, as well as providing a wide surface area for ion exchange. In surgeries such as cholecystectomy, simple mastectomy, and inguinal hernia repair, alginate swabs were found to significantly reduce blood loss and operation time [5].

Since the 1980s, the successful clinical applications have led to the development of a large number of alginate wound dressing products in the global wound management market. Table 6.1 shows the many types of alginate wound dressings available in the UK market.

6.6 Summary

Today, woundcare practitioners have at their disposal a wide range of functional wound dressings. These can be grouped under different product technologies, such as hydrocolloid dressings, alginate dressings, hydrogels, and foam dressings. Each of these product groups contains various brands that often have markedly different performance characteristics. As part of an overall wound management plan, these functional wound dressings can help facilitate fast wound healing by providing the optimal environment for healing to proceed. They are also able to deal with the odor, leakage, maceration, pain, infection, and other problems for wounded patients.

It should be pointed out that although wound dressings may not be able to significantly promote the healing process, the use of inappropriate dressings can often cause delayed wound healing, resulting in the complications of poor wound management, such as:

- delayed healing and/or wound deterioration
- increased risk of local or systemic infection
- increased demand for nursing time and increased dressing costs
- damage to the wound surface
- damage to the surrounding skin
- failure to control odor
- detrimental effect on the quality of patients' life

In order to provide the best solutions for woundcare practitioners, new types of advanced wound dressings and wound management materials are being developed to provide the means to ensure the best possible patient outcome. The aims of the new developments can be summarized into the following three areas:

- **E**fficacy of the material
- **E**ffectiveness of the product
- **E**fficiency of the treatment

These three "E"s symbolize the goal-oriented approach in the current and future development of wound management materials and products. As well as clinical efficacy and treatment effectiveness, cost is also an important issue in the development of new smart woundcare materials. In considering the cost of new treatment regimens, it is important to evaluate the cost not only in terms of direct treatment costs but also in terms of length of initial hospital stay, requirements for home care, additional bandaging regimens, and quality of the overall outcome. While new smart woundcare products are often more expensive than traditional products, in many cases this additional cost is justifiable. With respect to novel

Table 6.1 Alginate wound dressings listed in the British Drug Tariff.

Product name	Size	Price
ActivHeal Alginate	5 cm × 5 cm	58p
	10 cm × 10 cm	113p
	10 cm × 20 cm	278p
Algisite M	5 cm × 5 cm	87p
	10 cm × 10 cm	180p
	15 cm × 20 cm	484p
Algivon	5 cm × 5 cm	213p
	10 cm × 10 cm	359p
Algosteril	5 cm × 5 cm	87p
	10 cm × 10 cm	198p
	10 cm × 20 cm	334p
Curasorb	5 cm × 5 cm	70p
	10 cm × 10 cm	149p
	10 cm × 14 cm	241p
	10 cm × 20 cm	293p
	15 cm × 25 cm	515p
	30 cm × 61 cm	2703p
Curasorb Plus	10 cm × 10 cm	204p
Curasorb Zn	5 cm × 5 cm	80p
	10 cm × 10 cm	168p
	10 cm × 20 cm	330p
Kaltostat	5 cm × 5 cm	90p
	7.5 cm × 12 cm	196p
	10 cm × 20 cm	384p
	15 cm × 25 cm	661p
Medihoney Gel Sheet	5 cm × 5 cm	175p
	10 cm × 10 cm	420p
Melgisorb	5 cm × 5 cm	86p
	10 cm × 10 cm	179p
	10 cm × 20 cm	336p
Sorbalgon	5 cm × 5 cm	77p
	10 cm × 10 cm	162p
Sorbsan Flat	5 cm × 5 cm	80p
	10 cm × 10 cm	168p
	10 cm × 20 cm	315p
Suprasorb A	5 cm × 5 cm	58p
	10 cm × 10 cm	114p
Tegaderm Alginate	5 cm × 5 cm	78p
	10 cm × 10 cm	164p
Trionic	5 cm × 10 cm	119p
	10 cm × 15 cm	268p
	10 cm × 20 cm	332p

functional wound dressings such as those made from alginate fibers, it has been shown that these products are not only cost-effective, but they are also proven to be extremely beneficial in terms of their ability to reduce pain, odor, and leakage from the wounds.

References

1 Agboh, O.C. and Qin, Y. (1997). Chitin and chitosan fibers. *Polym Adv Technol* 8: 355–365.

2 Blaine, G. (1946). The use of plastics in surgery. *Lancet* ccli (2): 525–528.

3 Blaine, G. (1947). Experimental observations on absorbable alginate products in surgery. *Ann Surg* 125: 102–114.

4 Blaine, G. (1951). A comparative evaluation of absorbable haemostatics. *Postgrad Med J* 27: 613–620.

5 Blair, S.D., Jarvis, P., Salmon, M. et al. (1990). Clinical trial of calcium alginate haemostatic swabs. *Br J Surg* 77: 568–570.

6 Bray, C., Blaine, G., and Hudson, P. (1948). New treatment for burns, wounds and haemorrhage. *Nurs Mirror* 86: 239–242.

7 East, G.C. and Qin, Y. (1993). Wet-spinning of chitosan and the acetylation of chitosan fibers. *J Appl Polym Sci* 50: 1773–1779.

8 Fraser, R. and Gilchrist, T. (1983). Sorbsan calcium alginate fiber dressings in footcare. *Biomaterials* 4: 222–224.

9 Gilchrist, T. and Martin, A.M. (1983). Wound treatment with Sorbsan-an alginate fiber dressing. *Biomaterials* 4: 317–320.

10 Groves, A.R. and Lawrence, J.C. (1986). Alginate dressing as a donor site haemostat. *Ann R Coll Surg Engl* 68: 27–28.

11 Hinchley, H. and Murray, J. (1989). Calcium alginate dressings in community nursing. *Pract Nurs* 2: 264–268.

12 Hinman, C.D. and Maibach, H. (1963). Effect of air exposure and occlusion on experimental human skin wounds. *Nature* 200: 377.

13 Ho, J., Walsh, C., Yue, D. et al. (2017). Current advancements and strategies in tissue engineering for wound healing: a comprehensive review. *Adv Wound Care* 6 (6): 191–209.

14 Knecht, E. and Hibbert, E. (1926). Wet spinning of chitin fibers. *J Soc Dye Colour* 42: 343.

15 Kunike, G. (1926). Regenerated fibers from chitin. *J Soc Dye Colour* 42: 318.

16 McMullen, D. (1991). Clinical experience with a calcium alginate dressing. *Dermatol Nurs* 3 (4): 216–219.

17 Odugbesan, O. and Barnett, A.H. (1987). Use of a seaweed-based dressing in management of leg ulcers in diabetics: a case report. *Pract Diabet Int* 4: 46–47.

18 Oliver, L.C. and Blaine, G. (1950). Haemostasis with absorbable alginates in neurosurgical practice. *Br J Surg* 37: 307–310.

19 Passe, E.R.G. and Blaine, G. (1948). Alginates in endaural wound dressing. *Lancet* 2: 651.

20 Qin, Y. (2005). Superabsorbent cellulosic fibers for wound management. *Text Mag* 32 (1): 12–14.
21 Qin, Y. (2007). *Functional Wound Dressings*. Beijing: China Textile Press.
22 Qin, Y. (2007). Functional cellulosic fibers for wound management. *Chem Fibers Int* 1–2: 34–35.
23 Qin, Y. (2008). The preparation and characterization of chitosan wound dressings with different degrees of acetylation. *J Appl Polym Sci* 107 (2): 993–999.
24 Qin, Y. (2016). *Medical Textile Materials*. Cambridge: Woodhead Publishing.
25 Qin, Y. (2022). *Silver Containing Functional Wound Dressings*. Beijing: China Textile Press.
26 Qin, Y., Hu, H., and Luo, A. (2006). The conversion of calcium alginate fibers into alginic acid fibers and sodium alginate fibers. *J Appl Polym Sci* 101 (6): 4216–4221.
27 Thomas, S. (1985). Use of a calcium alginate dressing. *Pharm J* 235: 188–190.
28 Thomas, S. (1994). Wound care update. A structured approach to the selection of dressings. *Nurs RSA* 9 (4): 14–16.
29 Thomas, S. (1997). A guide to dressing selection. *J Wound Care* 6 (10): 479–482.
30 Turner, T.D. (1989). Development of wound dressings. *Wounds* 1 (3): 155–171.
31 Williams, C. (1994). Sorbsan. *Br J Nurs* 3: 677–680.
32 Winter, G.D. (1962). Formation of scab and the rate of epithelialization of superficial wounds in the skin of the young domestic pig. *Nature* 193: 293–294.

Further Reading

Bennett, G. and Moody, M. (1995). *Wound Care for Health Professionals*. London: Chapman and Hall.
Dealey, C. (1994). *The Care of Wounds*. Oxford: Blackwell Science Ltd.
Leaper, D.J. and Harding, K.G. (1998). *Wounds: Biology and Management*. Oxford: Oxford University Press.
Morison, M., Moffatt, C., Bridel-Nixon, J. et al. (1997). *Nursing Management of Chronic Wounds*. London: Mosby.
Thomas, S. (1990). *Wound Management and Dressings*. London: The Pharmaceutical Press.
Wardrope, J. and Smith, J.A.R. (1992). *The Management of Wounds and Burns*. Oxford: Oxford University Press.

7

Absorption and Interactive Properties of Alginate Wound Dressings

7.1 Introduction

Since Sorbsan™ calcium alginate wound dressing was first commercialized in 1981, there have been a large number of alginate wound dressings launched in the healthcare market, mainly for the treatment of medium to highly exuding wounds. There are now more than 10 types of alginate wound dressings available on the British Drug Tariff alone [1, 6, 18, 22]. These products differ in many respects. With the raw materials used for the production of fibers and wound dressings, alginate is chemically a polymeric acid composed of α-L-guluronic acid (G) and β-D-mannuronic acid (M). Alginate extracted from different types of seaweed can differ significantly in their G and M contents and also in the GG and MM contents. It is known that alginate high in G and GG contents can form more rigid gels than those high in M and MM contents, hence the differences in G and M content can have an important effect on the absorption and interactive properties of alginate fibers and wound dressings.

Alginate fibers were first made by extruding an aqueous solution of sodium alginate into an aqueous calcium chloride bath [17]. During the production process, the processing conditions can be altered to produce fibers with different calcium and sodium contents. It is known that fibers containing a small proportion of sodium ions are more absorbent than pure calcium alginate fibers [14, 15]. The production process can be engineered to make alginate fibers with required proportions of calcium and sodium ions.

Alginate fibers can be converted into wound dressings by using a number of textile processes. Due to its simplicity and also the high absorbency of the product, nonwoven is the main form of alginate wound dressings. A number of nonwoven structures such as needling, pressure rolling, and hydroentanglement have been used for making alginate wound dressings, resulting in products with significantly different absorption capacities.

In view of the diversified techniques that have been used for the production of alginate fibers and wound dressings, the absorption characteristics differ for alginate wound dressings made from different raw materials in different processing conditions, with polymer, fiber, and textile structures having a significant effect on the performances of various types of alginate wound dressings.

Alginate Fibers and Wound Dressings: Seaweed Derived Natural Therapy, First Edition. Yimin Qin.

This chapter summarizes the absorption and also interactive properties of alginate wound dressings.

7.2 Characterization Methods

Many types of alginate wound dressings are commercially available, offering a variety of different constructions and properties that are useful for the study of the absorption and interactive properties so as to correlate the chemical/physical structures of the alginate fibers and the performance of each dressing. Seven commercially available alginate dressings were used to characterize their absorption properties. Among them, Sorbsan (Maersk Medical) is a calcium alginate dressing made from alginate with a high mannuronate content. Tegagel™ (3 M) is made from the same type of fibers as Sorbsan but the fibers in Tegagel are hydro-entangled to form a nonwoven dressing, while Sorbsan is made in a loose form by pressurizing the carded web with pressure rollers. Curasorb™ (Kendall Healthcare) is a type of calcium alginate wound dressing with a high guluronate content, made by needle punching a nonwoven felt. Kaltostat™ (ConvaTec) is made with a similar type of alginate as Curasorb and is also a needled nonwoven felt, but in the Kaltostat fibers, the alginic acid is in a mixed form of calcium and sodium alginate, the sodium ions being introduced into the fibers during the production process in order to improve the absorbency and gel-forming ability of the dressing. In addition, Algosteril™ from Laboratory Brothier, Urgosorb™ from Urgo, and Tegagen HG™ from 3 M were also used.

Several non-alginate wound dressings were also evaluated, which include a knitted gauze made of viscose rayon continuous filament yarn (Smith & Nephew), a woven cotton gauze (ConvaTec), a nonwoven gauze made of cotton/polyester blend (Johnson & Johnson), and a polyester nonwoven felt (SSL International).

7.2.1 Test on Absorbency

Samples were cut to 5 cm × 5 cm sizes and conditioned at 25 °C and 65% relative humidity overnight. The dressings were then weighed (W) before being placed in plastic Petri dishes (90 mm in diameter) and wetted with 40 times their own weight of solution A, which is an aqueous solution containing 142 mmol of sodium chloride and 2.5 mmol of calcium chloride (British Pharmacopeia Monograph for Alginate Dressings and Packings, 1994). The dish was then placed in a 37 °C oven for 30 minutes. After that, the dressing was lifted out of the solution by holding with a forcep at one corner. The solution was left to drip for 30 s and the wet dressing was weighed (W_1). The sample was then placed in a centrifuge tube half filled with knitted gauze to contain the spin-off liquid. After centrifuging at 1200 rpm for 15 minutes, the dressing was taken out and weighed again (W_2). Finally, the dressing was dried to constant weight at 105 °C for four hours and the weight was weighed (W_3).

The absorption capacity of the dressing equals to $(W_1 - W)/W$ g/g. The fluid that is held within the dressing is divided into two parts, i.e. those held in the textile structure between the fibers and those held inside the individual fibers. In this experiment, $W_1 - W_2$ is the weight of fluid held between the fibers while $W_2 - W_3$ is the weight of fluid held within the fibers. The ratio of $(W_1 - W_2)/W_3$ and $(W_2 - W_3)/W_3$ are calculated to convert the fluid absorption into gram fluid absorbed per gram of dry fiber, which can be used to compare different types of dressings.

7.2.2 Fiber Calcium and Sodium Contents

The fiber calcium and sodium contents were analyzed by using atomic absorption spectroscopy. Fibers were first digested in concentrated sulfuric acid before tests were carried out. The calcium and sodium contents are expressed as the percentage of carboxylic acid groups being in calcium salt or sodium salt. This assumes that all the carboxylic acid groups on the alginate molecular chain exist either in calcium salt or in sodium salt. Assuming the wt/wt calcium content of the fiber is C_1 and the sodium content is C_2, the fiber calcium content equals to $[C_1/20]/[C_1/20 + C_2/23] \times 100\%$, while the fiber sodium content is $[C_2/23]/[C_1/20 + C_2/23] \times 100\%$ (note: one calcium ion binds with two carboxylic acid group).

7.2.3 Gel Swelling

The gel swelling abilities of the alginate fibers were measured by placing 0.2 g fiber in 100 ml of either distilled water or 0.9% wt/wt aqueous sodium chloride solution (normal saline). After one hour, the fibers were separated with the contacting solution and placed in a centrifuge tube with the bottom half filled with knitted viscose rayon fabric to contain the spin-off solution. The centrifuge was carried out at 1200 rev/minute for 15 minutes. After that, the fiber (W_1) was dried at 105 °C to constant weight (W_2). The gel-swelling ratio is expressed as the ratio between the weight of the wet sample and that of the dry sample, i.e. W_1/W_2.

7.2.4 Wet Integrity

The wet integrity test was carried out by placing a piece of 5 cm × 5 cm dressing in a wide necked 250 ml conical flask. After adding 50 ml of solution A, the flask was swirled gently for 60 s. The dressing was considered wet dispersible if the fibers were separated to leave no evidence of the original structure. It was considered wet integral if there was clear evidence of the original structure.

7.2.5 Wicking Behavior

The wicking behavior was tested on a gel blocking test kit, which was made of a semi-spheric plastic cup with a diameter of 50 mm and a depth of 5 mm. In total, 5 ml of 1.5% aqueous sodium citrate solution was first placed in the cup. A piece of 10 cm × 10 cm dressing was then placed on top of the solution. After one minute, the dressing was lifted up and the largest diameter on the wet circle of the dressing was measured. Wicking is expressed as the ratio between this diameter and the diameter of the original cup.

7.2.6 Dry and Wet Strength

The dry and wet strength of the alginate wound dressings were tested by cutting a strip of felt 2 cm in width from each direction of the felt. For the two directions, the strength for the weak direction was quoted. For the wet strength test, the felt was cut and then wetted with solution A similar to the absorbency test. After 30 minutes in a 37 °C oven, the felt was lifted out of the solution and tested for breaking strength. The gauge length was 50 mm and the cross-head speed was 50 mm/minute.

7.3 Absorption of Wound Fluid by Alginate-Based Wound Dressings

7.3.1 Absorption Mechanism of Alginate Wound Dressings

Alginate wound dressings have a unique absorption mechanism for wound fluid. During the production process for alginate fibers, aqueous solutions of sodium alginate are extruded via fine holes into coagulation baths usually containing calcium chloride. Upon contact with calcium ions, sodium alginate exchanges ions with calcium ions, and since calcium ion is a divalent metal ion capable of forming water-insoluble calcium alginate, alginate precipitates out in the form of a swollen gel. Upon further stretching, washing, and drying, calcium alginate fibers can be made. When in contact with wound exudate, a reverse process happens. As the calcium alginate fibers are in contact with fluid containing sodium ions, part of the calcium ions in the fiber exchange with the sodium ions in the contacting fluid. Since sodium alginate is water soluble, water is drawn into the fiber and the fiber swells to form a gel. This unique ion-exchange process is shown in Figure 7.1.

When alginate fibers are processed into nonwoven wound dressings, absorption of fluid by the dressing takes place in two forms, i.e. the fluid can either be absorbed into the fiber structure or absorbed between fibers in the textile structure. By centrifuging the wet dressings, it is possible to separate those liquid held between the fibers ($W_1 - W_2$) from the wet fibers (W_2) and by bone drying the wet fibers, it is possible to further separate the fluid held inside the fiber ($W_2 - W_3$) from the fiber material itself (W_3).

In the case of a wound dressing, the fact that fluid can be absorbed into the fiber structure is important in several respects. First, when fluid is absorbed into the fibers, they swell as a result. As the fibers expand upon swelling, the free spaces between the fibers are closed, and any bacteria that are in the textile structure or in the wound exudate are then immobilized. This action limits wound infection and cross-infection in a hospital ward. Second, if the fluid is held between the fibers by capillary forces, it can easily migrate along the textile structure. Clinically, this would mean the spreading of wound exudate from the wounded area to the surrounding healthy skins also covered by the dressing, hence causing skin maceration. Therefore, it is ideal that the dressing be capable of absorbing a large amount of exudate, and at the same time containing the fluid within the fiber structure.

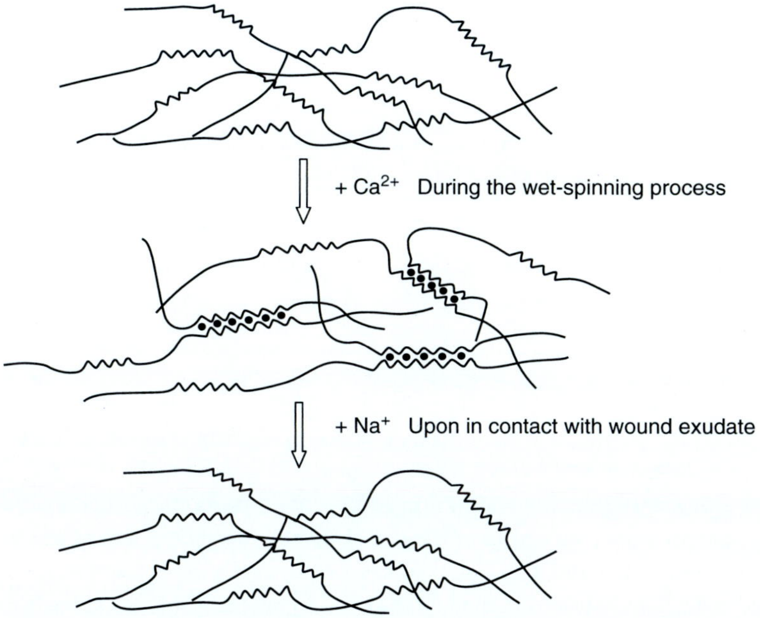

Figure 7.1 An illustration of the ion-exchange process of alginate during fiber formation and application.

7.3.2 Absorbency of the Various Types of Alginate Wound Dressings

In the absorption test described before, $(W_1 - W)/W$ reflects the amount of fluid retained by the wound dressings on a gram per gram basis. This test is the standard absorbency test for alginate wound dressings. As shown in Table 7.1, the various types of alginate wound dressings have considerably different absorption capabilities. In the case of alginate wound dressings, the needle-punched nonwoven felt made of calcium/sodium alginate (Kaltostat) had the highest absorbency of about 16.3 g/g, while the hydroentangled dressing made from high M calcium alginate fiber (Tegagel) has the lowest absorption capacity of about 4.8 g/g. The other two types of dressings, i.e. the needle-punched high G felt (Curasorb) and the pressure rolled un-needled high M felt (Sorbsan), have similar absorption capacities at about 12 g/g [14].

Table 7.1 Absorption properties of commercial alginate nonwoven dressings.

Product	$(W_1 - W)/W$	$(W_1 - W_2)/W_3$	$(W_2 - W_3)/W_3$	$(W_1 - W_2)/(W_2 - W_3)$
Sorbsan	11.98 ± 0.55	73.7 ± 2.11	15.2 ± 0.65	4.83 ± 0.25
Kaltostat	16.30 ± 0.63	47.1 ± 1.89	14.8 ± 0.71	3.18 ± 0.12
Curasorb	12.80 ± 0.53	32.0 ± 1.22	3.70 ± 0.13	8.62 ± 0.37
Tegagel	4.80 ± 0.16	18.6 ± 0.81	10.6 ± 0.39	1.75 ± 0.11

Table 7.2 Absorption properties of non-alginate products.

Product	$(W_1 - W)/W$	$(W_1 - W_2)/W_3$	$(W_2 - W_3)/W_3$	$(W_1 - W_2)/(W_2 - W_3)$
Knitted gauze of viscose rayon filament yarn	1.88 ± 0.081	4.64 ± 0.23	2.72 ± 0.13	1.71 ± 0.10
Woven gauze	4.83 ± 0.21	11.5 ± 0.48	6.45 ± 0.29	1.78 ± 0.11
Nonwoven gauze	8.90 ± 0.34	24.0 ± 1.11	2.15 ± 0.11	11.04 ± 0.42
Nonwoven polyester fabric	18.35 ± 0.81	67.6 ± 2.15	3.97 ± 0.16	17.01 ± 0.77

Alginate fibers can be made from different types of alginate having different guluronate and mannuronate contents, and the fibers can also be made into either calcium alginate or a mixed salt fiber containing different proportions of calcium and sodium ions. The nonwoven structure of the dressings can also be varied by adopting different nonwoven technologies, such as needle punching, pressure rolls, and hydroentanglement. The needle-punched felt has the highest level of porosity while in the hydroentangled felt, the individual fibers are heavily entangled and the pore structure is reduced. In general, the absorption capacities increase with the increase in the porosity of the dressings and the increase in the sodium content of the fibers.

Table 7.2 shows the absorption capacities of a number of non-alginate dressings. Knitted gauze made of viscose rayon filament yarn had the lowest absorption at about 1.88 g/g, while the nonwoven polyester fabric absorbs 18.3 g fluid per gram felt, the highest among all the dressings tested. For all the cellulosic type dressings, i.e. knitted viscose rayon gauze, woven cotton gauze, and nonwoven gauze, the absorption capacity increases with the increase in the porosity of the textile structure, in the order of knitted, woven, and nonwoven. Although alginate is a much more hydrophilic material than polyester, all the nonwoven alginate dressings tested have lower absorbency than the polyester nonwoven fabric. The hydrophobic nature of the polyester fibers made the nonwoven structure rigid when wet, which supports the porous structure of the dressing to hold a large amount of fluid.

7.3.3 Fluid Retention Between Fibers and Inside Fibers

By applying centrifuge to the wet dressing, the fluid that is retained between the fibers in the textile structure is separated from the original dressing structure. Furthermore, the fluid that is retained inside the fibers can be measured by drying it off the fibers. These techniques can be used to reveal the exact distribution of fluid retained by the textile structure and by the fiber itself.

As shown in Table 7.1 for the absorption properties of several types of alginate wound dressings, it is clear that for all the dressings, by far the majority of the fluid is held in the textile structure rather than in the fibers. However, the ratio of $(W_1 - W_2)/(W_2 - W_3)$ differs greatly for the different types of dressings, indicating that the distribution of fluid in the dressing varies for the different types of dressings. $(W_2 - W_3)/W_3$ reflects the amount of fluid held in the fiber per gram of dry fiber. The highest figure comes from the high M calcium alginate fiber, since it exchanges ions readily and gels rapidly on contact with sodium-containing liquid. However, although Sorbsan and Tegagel were made with similar high M fibers, the Sorbsan fiber had a higher gel-swelling ratio than the Tegagel fiber, indicating the effect of the nonwoven structure on the swelling of the fibers. It is also interesting to see the difference between Curasorb and Kaltostat. Although they are both made of high G-type alginate, the Kaltostat fiber contains about 80% calcium alginate and 20% sodium alginate, while the Curasorb is mostly calcium alginate. Kaltostat takes about 14.8 g solution A inside the fiber per gram of dry fiber, while Curasorb absorbs only 3.7 g.

In terms of the distribution of fluid, Sorbsan has a $(W_1 - W_2)/(W_2 - W_3)$ ratio of 4.83, while Tegagel is 1.75. The loose textile structure made Sorbsan more easy to retain fluid within the textile structure, hence its absorbency between fibers is 73.7 g per gram of dry fiber, while the corresponding number for Tegagel is only 18.6 g/g. Curasorb also has a much higher $(W_1 - W_2)/(W_2 - W_3)$ ratio than Kaltostat. Its rigid structure of high G calcium alginate made it less able to swell and less absorbing than the Kaltostat dressing.

7.3.4 A Comparison of Absorption Properties Between Alginate Felt and Rope

Compared to the felt dressings, alginate ropes are loosely assembled fibers with little fiber/fiber interaction. The main difference between the felt and rope is the accessibility of the individual fibers to the fluid, mainly because the ropes take a cylindrical form with the inner part of the rope further away from the fluid than in the felt products. Table 7.3 shows the absorption properties of Sorbsan ribbon and Kaltostat rope. In general, compared to the nonwoven dressings made of the same type of fibers, the differences in the various parameters are small.

7.3.5 Effect of Sterilization on the Absorption Properties of Alginate Dressings

Table 7.4 shows the absorption properties of sterilized and un-sterilized alginate nonwoven dressings and ropes. The sterilization is done by r-irradiation at 25 kGy. Post-sterilization, there is a significant reduction in the $(W_2 - W_3)/W_3$ parameter, indicating that the sterilization process has either made the alginate

Table 7.3 Absorption behavior of alginate ropes.

Product	$(W_1 - W)/W$	$(W_1 - W_2)/W_3$	$(W_2 - W_3)W_3$	$(W_1 - W_2)/(W_2 - W_3)$
Sorbsan ribbon	10.85 ± 0.51	52.5 ± 2.12	13.2 ± 0.61	3.97 ± 0.21
Kaltostat rope	11.46 ± 0.48	48.6 ± 1.65	15.6 ± 0.66	3.10 ± 0.14

Table 7.4 A comparison of the absorption properties of sterile and unsterile alginate dressings and ropes.

Product	$(W_1 - W)/W$	$(W_1 - W_2)/W_3$	$(W_2 - W_3)/W_3$	$\frac{(W_1 - W_2)}{(W_2 - W_3)}$
Un-sterilized alginate nonwoven dressing	20.85 ± 0.85	82.4 ± 2.98	24.3 ± 0.92	3.38 ± 0.14
Sterilized alginate nonwoven dressing	18.72 ± 0.82	82.6 ± 3.12	21.0 ± 0.87	3.93 ± 0.14
Un-sterilized alginate rope	19.52 ± 0.95	76.3 ± 2.41	23.7 ± 1.11	3.22 ± 0.14
Sterilized alginate rope	16.57 ± 0.58	75.3 ± 2.65	17.4 ± 0.67	4.32 ± 0.21

Note: the alginate fibers used in the above experiment are a mixed calcium/sodium alginate containing about 70% calcium alginate and 30% sodium alginate.

more crystallized or cross-linked in some form, which makes it less absorbent. Interestingly, in both the nonwoven dressing and rope, the absorption between fibers $(W_1 - W_2)/W_3$ remains largely unchanged before and after sterilization, which means that the sterilization process has little effect on the textile structure. Since the fibers absorb less after sterilization, the overall absorption capacities of the dressings were reduced post-sterilization.

7.3.6 Effect of Guluronate and Mannuronate Contents

Table 7.5 shows the analysis results for Algosteril, Curasorb· and Sorbsan dressings. All three dressings are made of calcium alginate fibers. Both Algosteril and Curasorb are made of high G alginate, while Sorbsan is made of high M alginate. The high G and high M type fibers differ significantly in the gel-swelling ratio in saline. As can be seen in Table 7.5, Sorbsan had a significantly higher saline swelling ratio than both Algosteril and Curasorb, while the water-swelling ratios were similar for the three types of alginate fibers. This is understandable since the calcium ions are bound more strongly with high G alginate, hence ion exchange with sodium ions in the solution is more difficult for high G alginate fiber. For high M fibers, the calcium

Table 7.5 Performance characteristics of three calcium alginate dressings.

Test criteria	Algosteril	Curasorb	Sorbsan
Weight per unit area (g/m^2)	125.55 ± 8.15	127.50 ± 5.80	125.20 ± 6.50
Absorbency (g/g)	14.27 ± 0.41	14.77 ± 0.36	16.75 ± 0.27
Wet integrity	Integral	Integral	Dispersible
Gel swelling ratio in water (g/g)	1.85 ± 0.13	2.12 ± 0.12	2.14 ± 0.14
Gel swelling ratio in saline (g/g)	5.23 ± 0.42	4.42 ± 0.13	13.91 ± 0.77
Fiber diameter (μm)	25.25 ± 1.22	15.27 ± 1.68	17.4 ± 1.87

ions in the fibers can be easily replaced by sodium ions in the solution, resulting in a better gelling ability. As the fibers can swell better, the Sorbsan dressings can absorb more fluid than the Algosteril and Curasorb dressings under the same test conditions. Because they swell less, Algosteril and Curasorb dressings have better-wet integrity than Sorbsan.

7.3.7 Effect of Calcium and Sodium Contents

Since sodium alginate is water soluble, by making alginate fibers a mixture of calcium and sodium alginate, it is possible to improve the absorbency of the alginate dressings. Both Tegagen HG and Kaltostat are made of calcium and sodium alginates. The former is a high M alginate, while the latter is made of high G alginate. Analysis results showed that the proportions of alginic acid as calcium and sodium salts were, respectively, about 65/35 and 80/20 for Tegagen HG and Kaltostat, as compared to 96.6/3.4 and 99.6/0.4 for Sorbsan and Algosteril, respectively. As can be seen in Table 7.6, both Tegagen HG and Kaltostat have significantly higher water-swelling ratios than Sorbsan and Algosteril, indicating the high water-holding abilities of the sodium ions in the fibers. Due to the ion strength in the solution, the saline-swelling ratios for Tegagen HG and Kaltostat were lower than the water swelling ratios, however, they were still much higher than the corresponding figures for Sorbsan and Algosteril.

Tegagen HG had an absorbency of 20.51 g/g, about 22% higher than Sorbsan under similar testing conditions. The absorbency of Kaltostat showed a similar level of improvement over Algosteril, although both fibers were made of similar high G alginate. With the calcium/sodium alginate wound dressings, it is interesting to note that the individual fibers within the dressing can have different proportions of calcium and sodium ions, as illustrated in the different degrees of swelling among individual fibers shown in Figure 7.2.

7.3.8 Effect of Nonwoven Structures

The Tegagel and Sorbsan dressings are made of similar high M calcium alginate fibers. Test results showed that the absorbency for Tegagel is only 4.68 g/g, which

Table 7.6 Performance characteristics of calcium/sodium alginate dressings.

Test criteria	Tegagen HG	Kaltostat
Fiber sodium content	~35%	~20%
Weight per unit area (g/m^2)	119.01 ± 15.11	120.5 ± 11.5
Absorbency (g/g)	20.51 ± 0.66	17.40 ± 0.35
Wet integrity	Dispersible	Not dispersible
Gel swelling ratio in water	37.67 ± 1.25	7.65 ± 0.35
Gel swelling ratio in saline	16.75 ± 0.85	5.79 ± 0.21
Fiber diameter (μm)	12.15 ± 0.65	16.3 ± 1.95

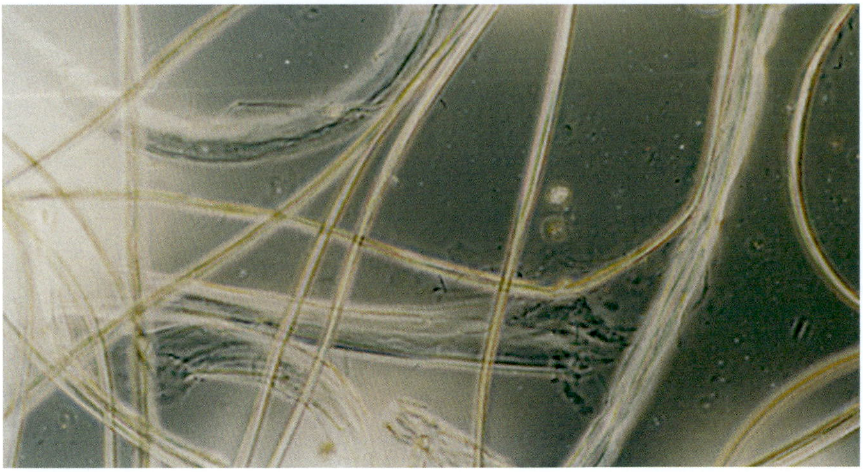

Figure 7.2 An illustration of the swelling in water of fibers in a calcium/sodium alginate wound dressing, ×200.

compares to 16.75 g/g for Sorbsan. As Tegagel is made of a hydroentangled structure, the fibers in the dressing are densely compressed, hence it is more difficult for the fluid to diffuse into the nonwoven structure. On the other hand, Sorbsan is made of an unneedled pressure-rolled structure, where the swelling of fibers is easy when wet. Because the fibers are closely adhered to each other, the Tegagel sample showed a high degree of wet integrity.

7.3.9 Effect of Adding CMC Into the Alginate Fibers

Because they are both water-soluble polysaccharides, sodium alginate and sodium carboxymethyl cellulose (CMC) can be mixed together in an aqueous solution and made into a composite fiber. The CMC component disrupts the regular structures of alginate and hence it is easy for the alginate/CMC fiber to swell in the presence of saline. As can be seen in Table 7.7, Urgosorb, which is an alginate/CMC composite fiber containing about 15% CMC, showed a significantly higher saline-swelling ratio than both Algosteril and Curasorb, though the three samples are made of similar high G calcium alginate. The absorbency for Urgosorb is more than 30% higher than Algosteril and Curasorb under similar test conditions.

Table 7.7 Comparison of alginate/CMC to other high G alginate dressings.

Test criteria	Urgosorb	Algosteril	Curasorb	Kaltostat
Absorbency (g/g)	20.35 ± 0.75	14.27 ± 0.41	14.77 ± 0.36	17.40 ± 0.35
Gel swelling ratio in water (g/g)	4.25 ± 0.25	1.85 ± 0.13	2.12 ± 0.12	7.65 ± 0.35
Gel swelling ratio in saline (g/g)	9.65 ± 1.42	5.23 ± 0.42	4.42 ± 0.13	5.79 ± 0.21

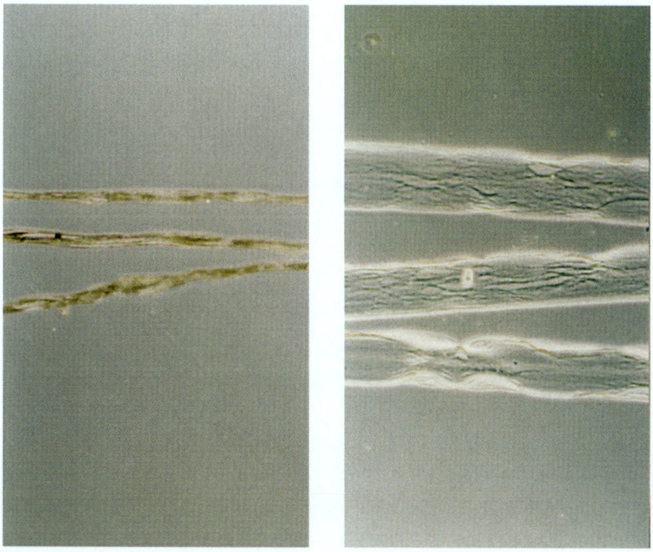

(a) (b)

Figure 7.3 An illustration of the swelling of alginate/CMC fibers, (a) dry; (b) wet in saline, ×200.

It is clear that by adding CMC into the fiber, the performances of the alginate dressings can be improved in a similar way as when sodium ions are introduced to the fibers. Since CMC and alginate are mixed in the spinning solution, the CMC component can be uniformly dispersed in the alginate fibers. Figure 7.3 shows the photomicrograph of alginate/CMC fibers when wet in saline, showing a high degree of swelling.

7.3.10 Wicking of Fluid

Alginate wound dressings are known to have the "gel blocking" properties whereby the fluid absorbed by the dressing does not spread along the fabric structure. Lateral wicking of fluid is difficult because the alginate fibers swell when in contact with wound exudate, thereby blocking pores in the nonwoven structure responsible for the passing of fluid. Figure 7.4 shows the photographs of fluid wicking when Sorbsan, Kaltostat, and Urgosorb dressings were placed in contact with 1.5% aqueous sodium citrate solution in a semi-spheric plastic cup with a diameter of 50 mm and a depth of 5 mm. For the Sorbsan dressing, which has a loose nonwoven structure, although it quickly turned into a piece of soft gel, the fluid was able to spread along the dressing to form a wet radius about 1.3 times the diameter of the original cup. The Kaltostat fibers were more difficult to gel and the wet radius was 1.5 times of the cup diameter, indicating the relatively poor gel-blocking properties of the dressing. Urgosorb showed the best gel-blocking properties, with the fluid virtually remaining on top of the plastic cup.

The "gel blocking" properties of alginate wound dressings can be explained in Figure 7.5, which shows the different absorption mechanisms for alginate fibers and

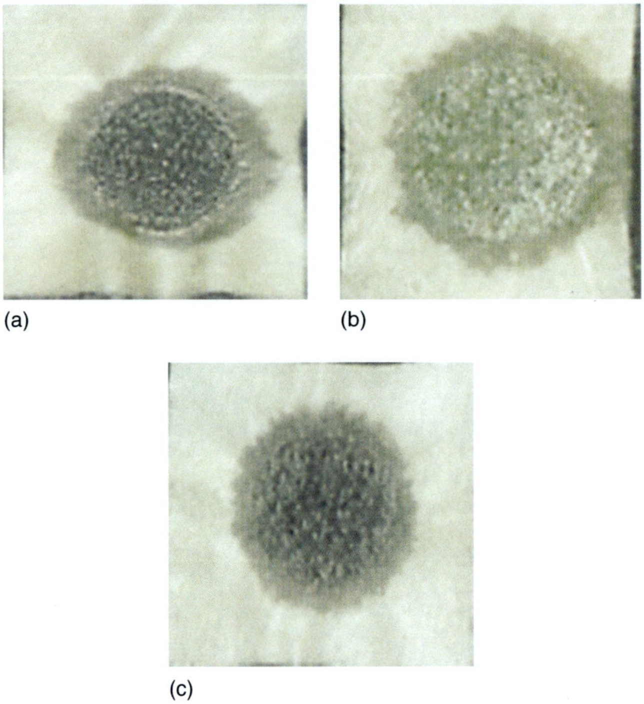

Figure 7.4 Wicking behavior for (a) Sorbsan; (b) Kaltostat, and (c) Urgosorb.

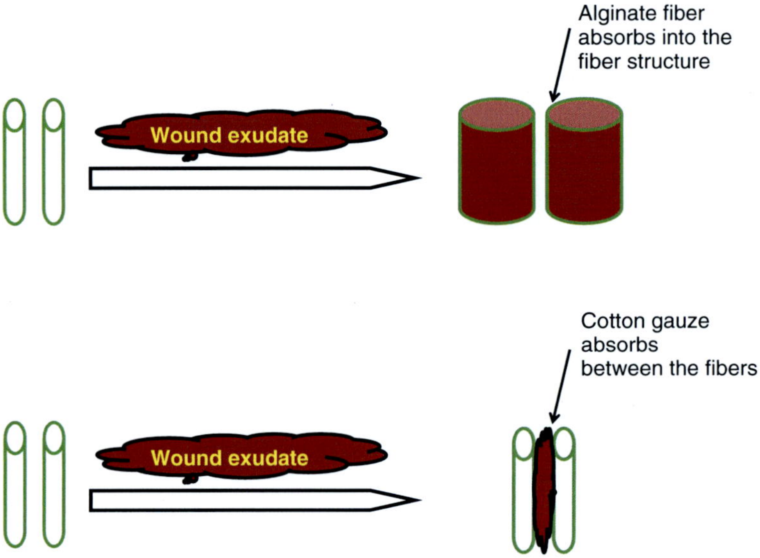

Figure 7.5 An illustration of the absorption mechanism for alginate fibers and cotton gauze.

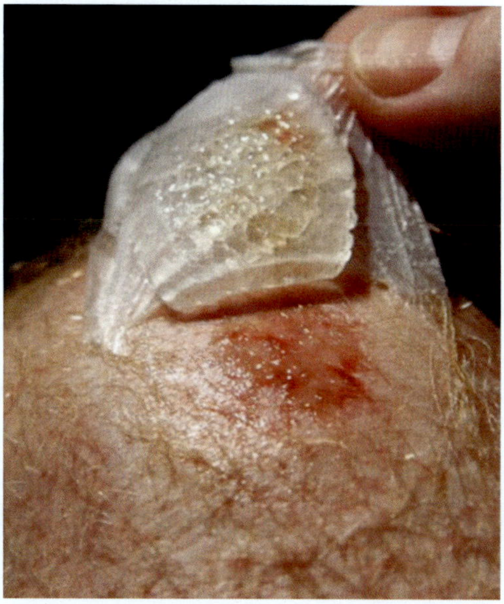

Figure 7.6 An illustration of the in-situ formation of alginate hydrogel on the wound surface.

cotton gauze. Alginate fiber absorbs a large quantity of fluid into the fiber structure through its unique ion exchange between calcium ions in the fiber and sodium ions in the contact solution, resulting in the swelling of the fiber which blocks the inter-fiber space that acts as the channel for the spreading of liquid. On the other hand, for cotton gauze, since the fiber itself barely swell in the presence of liquid, absorption takes place mainly in the capillary spaces between the fibers in the gauze, which can easily migrate along the gauze. The absorption of fluid into the fiber structure also helps to convert the alginate wound dressing into a piece of hydrogel when applied onto exuding wounds, as illustrated in Figure 7.6.

7.3.11 Dry and Wet Strength

Table 7.8 shows the dry and wet strength of Sorbsan, Kaltostat, and Urgosorb dressings. Because the Sorbsan dressings are made of loosely assembled fibers in a pressure-rolled structure, the dry strength of the Sorbsan dressing is considerably lower than the other two dressings, which are both made of needled nonwoven structures. When placed in solution A, all three types of alginate dressings had increased strength, most likely due to the adhesion of individual fibers when wet.

Table 7.8 Test results on the strength of alginate/CMC and other alginate dressings.

Test criteria	Urgosorb (Alginate/CMC)	Sorbsan	Kaltostat
Dry strength (N)	0.38 ± 0.06	0.13 ± 0.05	0.49 ± 0.07
Wet strength (N)	1.17 ± 0.18	0.81 ± 0.06	1.51 ± 0.41

Due to the high G nature of the alginate, the wet strength for the Kaltostat dressing is the highest among the three dressings. The Urgosorb dressing had a lower wet strength than Kaltostat, due most likely to the disruption of the regular structure by the presence of CMC within the alginate fiber.

7.4 Interactive Properties of Alginate Wound Dressings

7.4.1 Interactive Moisture Handling Properties of Alginate Wound Dressings

The fluid-handling capacity of alginate wound dressings ranges from about 15 to 25 g/100 cm^2. Under a compression bandage, this may be reduced to 5–10 g/100 cm^2. Therefore, a standard 10 cm × 10 cm dressing would not be able to cope with the exudate produced from a 20 cm^2 wound for more than about 12 hours. A standard polyurethane film dressing, which has a maximum moisture-vapor transmission rate (MVTR) of approximately 1000 g/m^2/24 hours [20], applied over an alginate sheet, allows for the loss of another 5 g of fluid in 24 hours. This makes a total fluid-handling capacity for an alginate/film dressing combination of 10–15 g/100 cm^2 on the first day, which is still less than the volume of fluid produced by a heavily exuding leg ulcer, or donor site. On the second day, however, the alginate dressing, being fully saturated, would be unable to absorb any more exudate, so fluid would rapidly accumulate beneath the dressing, resulting in leakage and/or maceration of the surrounding skin [13].

The application of an "intelligent" film dressing over a sheet of alginate might partially resolve such problems. In the presence of liquid, such films have an MVTR of 5000–10 000 g/m^2/24 hours, which greatly enhances their ability to cope with exudate production. As the wound begins to dry, the MVTR of the film would decrease and thus help to preserve moisture in the alginate fibers. Such a combination could prove particularly valuable in the treatment of donor sites.

The main advantages of alginate dressings are their ability to form a moist environment on the wound surface that facilitates optimal wound healing and permits pain- and trauma-free removal. Therefore, they should be regarded as low-adherent and gel-forming interface layers, rather than as absorbent dressings in their own right. This description emphasizes the importance of an appropriate secondary dressing to control moisture-vapor loss and provide a bacterial barrier function.

7.4.2 Biologically Interactive Properties of Alginate Wound Dressings

The biologically interactive properties of alginate wound dressings are first illustrated from the different healing properties of alginate wound dressings with different chemical compositions. In 1992, a survey was conducted into the management of fungating wounds and radiation-damaged skin by specialist centers throughout the UK [19]. Although the 114 respondents rated Sorbsan and Kaltostat as equivalent in terms of fluid-handling properties, Sorbsan was considered to be much superior to Kaltostat in the treatment of malodorous, necrotic, or infected wounds. It was pointed out that the relatively small differences in the structure of the alginate dressings may have important implications for the way in which they perform at a cellular level within the wound. In particular, the interaction between the alginate molecule and macrophage cells plays a key role in many physiological and pathophysiological processes through synthesizing various biologically active molecules called cytokines.

A major cytokine secreted by macrophages is tumor necrosis factor (TNF)α, also known as cachectin, which is produced when the cells are exposed to endotoxins (lipopolysaccharide [LPS] molecules derived from bacterial cell walls). It was first described as a tumor cytotoxic agent, having cytotoxic properties against both tumor cells and normal cells infected with intracellular pathogens. It is also a very important inflammatory mediator, which modulates many physiological and immunological functions and has been implicated in inflammatory conditions, such as rheumatoid arthritis, Crohn's disease, multiple sclerosis, and the cachexia associated with cancer or human immunodeficiency virus infection.

Experimentally it has been shown that the production of endotoxin-induced TNFα inhibits the effect of growth factors in the area of a wound, resulting in decreased collagen production and eventually impairment of the healing process [8]. A reduction in collagen production has similarly been shown to result from the direct application of TNFα to human and animal fibroblasts *in vitro* [16]. Paradoxically, however, it has been suggested that the ability of TNFα to inhibit collagen formation may be beneficial in fetal wounds, where it will limit fibroplasia and thus reduce scarring [3].

The ability of a macrophage to function in this way depends on the successful completion of the differentiation pathway of immature precursor cells to the mature macrophage. Circulating blood monocytes emigrate into extravascular tissue either to become resident organ-specific mature macrophages or to be recruited as immune effector cells at sites of inflammation, injury, allograft, or tumor rejection. Macrophages are the main cell type that regulates the wound healing cascade, and their deactivation halts the healing process. Wound macrophages can be stimulated (activated) by both endogenous and exogenous factors including alginate. It is interesting to note that high M alginate is more active than high G alginate in stimulating macrophage activities, hence can more effectively promote the healing of chronic ulcerative wounds.

7.4.3 Enzyme Inhibition Properties of Alginate Wound Dressings

Wound healing is a process of tissue repair, including various stages such as inflammation, cell proliferation, epithelization, and tissue remodeling. It is a complex process involving chemotaxis, phagocytosis, collagen formation and degradation, collagen remodeling, angiogenesis, and many other cell activities, where many enzymes play vital functions. From the beginning of wound formation to the completion of the healing process, many varieties of matrix metalloproteinases (MMPs) and their inhibitors are closely involved. During the wound-healing process, the main functions of MMPs include:

(1) Remove deactivated tissue;
(2) Regulate the interaction between epidermis and mesenchymal cells during keratinocyte migration;
(3) Participate in angiogenesis;
(4) Participate in the reconstruction of newly formed connective tissue; and
(5) Regulate growth factor activity.

The increased expression of MMPs and enhanced enzyme activity can lead to excessive degradation of extracellular matrix (ECM), such as in the case of chronic wounds, while underexpression and excessive inhibition of enzyme activity can lead to a large accumulation of ECM, which is manifested as hyperplasia of scar and keloid. It has been shown that the expression of MMP-1, MMP-2, MMP-8, and MMP-9 increased in the exudate of chronic ulcer wound [26, 27], which correspond to increased tissue degradation. At the same time, chronic wounds also showed very low levels of tissue inhibitors of matrix metalloproteinases (TIMPs), such as TIMP-1 and TIMP-2. These TIMPs have inhibitory activity against excessive levels of MMPs in chronic wounds and can have a promotional role in wound healing [23]. In the healing of chronic wounds, the balance between MMPs and TIMPs is important to restore the normal healing process and to promote the regeneration of extracellular matrix.

An important feature of chronic wounds is the persistence of inflammatory responses such as leukocytes and neutrophils, as well as the secretion of proteases and pro-inflammatory cytokines by macrophages. Ongoing inflammatory processes may explain the high level of MMPs in chronic wounds, resulting in sustained destruction of the extracellular matrix [12]. In order to promote chronic wound healing, it is necessary to use exogenous inhibitors to inhibit the activity of MMPs on the wound. In one trial, doxycycline, which can inhibit MMPs, was used in the care of diabetic foot ulcers [4]. Caveolin-1 also plays a role in the regulation of the MMP-1 gene and can be used as a drug in chronic wound care [7, 21].

Among the many types of MMPs involved in the wound healing process, MMP-9 is a zinc-containing endopeptidase that can degrade extracellular matrix, promote cell migration and angiogenesis, and participate in every stage of the wound healing process. Under normal conditions, MMP-9 is rarely expressed. When skin injury occurs, MMP-9 is activated, and as the wound-healing process proceeds, the level of MMP-9 expression decreases significantly [2]. Many studies have shown that the level of MMP-9 expression in chronic wound exudate is significantly higher than that in acute wound exudate. This over-expression can lead to an excessive proteolytic environment that extends the healing process from acute to chronic wounds [10].

TIMP-1 is a soluble glycoprotein mainly secreted by macrophages and connective tissue cells. It is widely distributed in the body through the catalytic region of MMP-9 with 1 : 1 reversible binding. TIMP-1 can inhibit the activities of MMP-9, and the ratio between MMP-9 and TIMP-1 is important in the wound healing process, playing vital roles in cell migration, extracellular matrix degradation, and remodeling [11]. Under normal physiological conditions, MMP-9 and TIMP-1 maintain a dynamic balance, coordinate the degradation and reconstruction of extracellular matrix, and maintain the integrity of tissue structure and the stability of internal environment. When the ratio of MMP-9/TIMP-1 is high, the degree of excessive degradation is heavy. Experimental results showed that the level of MMP-9 expression in diabetic foot wounds with poor healing condition was significantly higher than that in diabetic foot wounds with good healing condition. High levels of MMP-9 correspond to the existence of inflammation and poor wound healing. Ladwig et al. [9] found that the ratio of MMP-9/TIMP-1 was negatively correlated with the healing of pressure sores, and the ratio of MMP-9/TIMP-1 showed a decreasing trend with the healing of pressure sores.

Through its strong binding with zinc ions in the enzyme, alginate acts as an enzyme inhibitor during the wound healing process and can inhibit MMP-9 activities similar to TIMPs. It was found that calcium alginate can accelerate diabetic wound healing by inhibiting the activity of MMP-9 [24]. Wang et al. [25] studied the effect of calcium alginate dressing on MMP-9 in diabetic wound tissue. A round wound with a diameter of 20 mm was cut on the back of 36 male diabetic rats after anesthesia with aseptic surgical scissors. The experimental group (n = 18) was treated with calcium alginate dressing, and the control group (n = 18) was treated with cotton gauze. Western Blotting test results showed that the protein expression of MMP-9 in the experimental group was significantly lower than that of the control group at 3, 7, and 14 days after surgery, indicating that calcium alginate dressing could reduce collagen degradation and promote wound healing by inhibiting the expression of MMP-9. Figure 7.7 shows a comparison of wound healing for two groups of rats post-surgery.

Dinh et al. [5] found that the level of MMP-9 mRNA in normal skin was very low, but increased significantly at the time of injury, reaching a peak at 24 hours

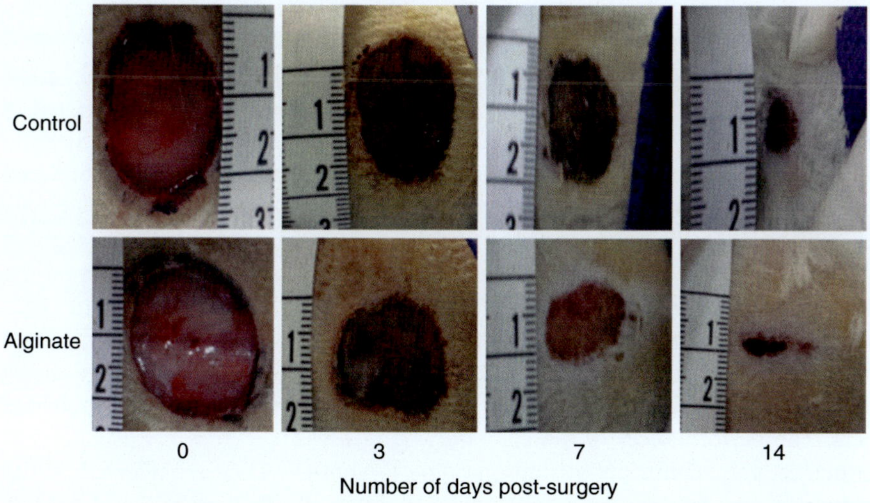

Figure 7.7 A comparison of wound healing for two groups of rats. Source: Adapted with permission from Wang [25].

before decreasing gradually. When the wound was completely re-epithelialized, MMP-9 returned to normal level. Wang et al. [25] showed that significant statistical differences in thickness of granulation were found in two groups of wounds covered with calcium alginate dressing and cotton gauze, respectively. At day 3, the thickness was (1476 ± 118) μm for the alginate group, while it was (812 ± 136) μm for the gauze group. At the same time, the level of collagen in control group was much lower than those in experimental group. At day 7, the level of collagen in the gauze group was $(15 \pm 5\%)$, while it was $(51 \pm 5\%)$ in the alginate group. Experimental results also show that the mRNA expression of MMP-9 in the alginate group decreased significantly when compared with those in the gauze group at days 3, 7, and 14 (0.028 ± 0.003 vs 0.037 ± 0.004, 0.026 ± 0.003 vs 0.033 ± 0.003, 0.021 ± 0.003 vs 0.030 ± 0.002). It was concluded that calcium alginate dressing promotes skin ulcer healing by inhibiting the expression of MMP-9 so as to reduce the degradation of collagen in diabetic rats.

Similar results were reported by Yao et al. [28]. After calcium alginate hydrogel containing transforming growth factor TGF-β3 was applied to human skin wounds, it was found that the level of MMP-9 was reduced in the experimental group, while the level of type I collagen was increased. It is believed that the calcium alginate hydrogel containing TGF-β3 can reduce the level of MMP-9 in the wound, increase the synthesis of type I collagen, and promote wound healing. But whether the effect was caused by alginate dressing or the role played by TGF-β3 remains unclear. In the study of Wang et al. [25], alginate dressing was directly used to treat diabetic wounds without interference from other factors, which confirmed its efficacy in inhibiting the expression of MMP-9.

7.5 Summary

Alginate is a bioactive polymer and the ability of alginate wound dressings to promote wound healing depends on its physical/chemical properties as well as its interactive properties with the wound. In terms of their physical/chemical properties, the absorption performances of alginate wound dressings are affected by a number of factors. Analysis results showed that a significant part of the absorption takes place inside the fiber structure, in addition to those liquid held between fibers in the textile structure. High M alginate and high G calcium/sodium alginate fibers absorb more fluid into the fiber than high G calcium alginate fiber, resulting in a better gelling ability. The textile structure also has a significant effect on the absorption properties of alginate dressings, especially with regard to the liquid held between fibers. In general, needle-punched felt absorbs more than hydroentangled ones. The study on the effect of sterilization shows that r-irradiation mainly affects the absorption within the fibers but has little effect on the absorption between fibers. The performances of the alginate wound dressings currently available on the market vary significantly from one type to another. They are affected by the guluronate and mannuronate contents of the alginate, the calcium and sodium contents of the fiber, the nonwoven structure, and the additives into alginate. By introducing sodium ions or CMC into the alginate fibers, major improvements can be made on the absorbency of the alginate wound dressings. In addition to the many physical/chemical properties, the G/M content of alginate also has an important effect on the bioactivities of the alginate dressing, with high M alginate being more effective in stimulating macrophage activities.

References

1 Attwood, A.I. (1989). Calcium alginate dressing accelerate split graft donor site healing. *Br. J. Plast. Surg.* 42: 373–379.

2 Bellayr, I., Holden, K., Mu, X. et al. (2013). Matrix metalloproteinase inhibition negatively affects muscle stem cell behavior. *Int. J. Clin. Exp. Pathol.* 6 (2): 124–141.

3 Boyce, D.E., Thomas, A., and Hart, J. (1997). Hyaluronic acid induces tumour necrosis factor-alpha production by human macrophages in vitro. *Br. J. Plast. Surg.* 50 (5): 362–368.

4 Chin, G.A., Thigpin, T.G., Perrin, K.J. et al. (2003). Treatment of chronic ulcers in diabetic patients with a topical metalloproteinase inhibitor, doxycycline. *Wounds* 15: 315–323.

5 Dinh, T., Tecilazich, F., Kafanas, A. et al. (2012). Mechanisms involved in the development and healing of diabetic foot ulceration. *Diabetes* 61: 2937–2947.

6 Groves, A.R. and Lawrence, J.C. (1986). Alginate on donor site. *Ann. R. Coll. Surg. Engl.* 68: 27–28.

7 Haines, P., Samuel, G.H., Cohen, H. et al. (2011). Caveolin-1 is a negative regulator of MMP-1 gene expression in human dermal fibroblasts via inhibition of Erk1/2/Ets1 signaling pathway. *J. Dermatol. Sci.* 64: 210–216.

8 Kawaguchi, H., Hizuta, A., and Tanaka, N. (1995). Role of endotoxin in wound healing impairment. *Res. Commun. Mol. Pathol. Pharmacol.* 89 (3): 317–327.

9 Ladwig, G.P., Robson, M.C., Liu, R. et al. (2002). Ratios of activated matrix metallopmteinase-9 to tissue inhibitor of matrix metalloproteinase-l in wound fluids are inversely correlated with healing of pressure ulcers. *Wound Repair Regen.* 10 (1): 26–37.

10 Lazaro, J.L., Izzo, V., Meaume, S. et al. (2016). Elevated levels of matrix metalloproteinases and chronic wound healing: an updated review of clinical evidence. *J. Wound Care* 25 (5): 277–287.

11 Lobmann, R., Ambrosch, A., Schultz, G. et al. (2002). Expression of matrix-metalloproteinases and their inhibitors in the wounds of diabetic and non-diabetic patients. *Diabetologia* 45 (7): 1011–1016.

12 Menke, N.B., Ward, K.R., Witten, T.M. et al. (2007). Impaired wound healing. *Clin. Dermatol.* 25: 19–25.

13 Moody, M. (1991). Calcium alginate: a dressing trial. *Nurs. Stand. Spec. Suppl.* 13: 3–6.

14 Qin, Y. (2004a). Absorption characteristics of alginate wound dressings. *J. Appl. Polym. Sci.* 91 (2): 953–957.

15 Qin, Y. (2004b). Gel swelling properties of alginate fibers. *J. Appl. Polym. Sci.* 91 (3): 1641–1645.

16 Rapala, K.T., Vaha-Kreula, M.O., and Heino, J.J. (1996). Tumour necrosis factor-alpha inhibits collagen synthesis in human and rat granulation tissue fibroblasts. *Experientia* 51 (1): 70–74.

17 Speakman, J.B. and Chamberlain, N.H. (1944). The production of fibers from alginate. *J. Soc. Dyers Colour.* 60: 264–272.

18 Thomas, S. (1989). Sorbsan in leg ulcer. *Pharm. J.* 243: 706–709.

19 Thomas, S. (1992). *Current Practices in the Management of Fungsating Lesions and Radiation Damaged Skin*. Bridgend: SMTL.

20 Thomas, S., Loveless, P., and Hay, N.P. (1988). Comparative review of the properties of six semipermeable film dressings. *Pharm. J.* 240: 785–789.

21 Tu, G., Xu, W., Huang, H. et al. (2008). Progress in the development of matrix metalloproteinase inhibitors. *Curr. Med. Chem.* 15: 1388–1395.

22 Turner, T. (1989). The development of wound management products. *Wounds* 1: 155–171.

23 Vaalamo, M., Leivo, T., and Saarialho-Kere, U. (1999). Differential expression of tissue inhibitors of metalloproteinases (TIMP-1, -2, -3, and -4) in normal and aberrant wound healing. *Hum. Pathol.* 30: 795–802.

24 Wang, T., Liu, F., Gu, Q. et al. (2014). Histological study on effect of calcium alginate dressing on wound healing in rats. *Inf. Inflamm. Repair* 15 (3): 154–157.

25 Wang, T., Zhao, J., Mei, J. et al. (2016). Effects of calcium alginate dressings on expression of matrix metalloproteinases-9 in the wound tissue of diabetic rats. *Chin. J. Diabetes Mellit.* 8 (3): 162–167.

26 Wysocki, A.B., Staiano-Coico, L., and Grinnell, F. (1993). Wound fluid from chronic leg ulcers contains elevated levels of metalloproteinases MMP-2 and MMP-9. *J. Invest. Dermatol.* 101: 64–68.

27 Yager, D.R., Zhang, L.Y., Liang, H.X. et al. (1996). Wound fluids from human pressure ulcers contain elevated matrix metalloproteinase levels and activity compared to surgical wound fluids. *J. Invest. Dermatol.* 107: 743–748.

28 Yao, Y., Zhang, F., Zhou, R. et al. (2010). Continuous supply of TGFβ3 via adenoviral vector promotes type I collagen and viability of fibroblasts in alginate hydrogel. *J. Tissue Eng. Regener. Med.* 4 (7): 497–504.

8

Clinical Applications of Alginate Wound Dressings

8.1 Introduction

Alginate wound dressings are a type of novel biomedical material made from a marine biopolymer. After many years of development, there are now many types of alginate wound dressings in the international market, each with unique properties and applications. Taking into consideration of the latest results of clinical researches conducted around the world, this chapter summarizes the unique properties of alginate wound dressings including their "gel blocking" properties and their ability to promote wound healing, facilitate hemostasis, reduce pain, suppress bacteria growth, and lower treatment cost. Because of their high absorption capacity and gel-forming ability, alginate wound dressings are suited for the treatment of pressure sores, leg ulcers, diabetic foot ulcers, burn wounds, surgical wounds, and many other types of wounds with high levels of wound exudate.

8.2 Biocompatibility and Bioactivities of Alginate Wound Dressings

Many studies have confirmed that alginate wound dressings have good biocompatibility. In a comparative study of the properties of three hemostatic agents used in surgical practice, Blair et al. [9] found the calcium/sodium alginate wound dressing Kaltostat to be more effective than either oxidized cellulose or porcine collagen in controlling bleeding from a surgically inflicted wound in rabbit liver. In addition, alginate showed no tendency to cause intestinal obstruction when implanted into mesentery. In another study, Lansdown and Payne [42] implanted samples of Kaltostat subcutaneously in rats to evaluate their biodegradability and ability to evoke local tissue reactions. Implant sites were evaluated after 24 hours, 7 days, 28 days, and 12 weeks. Histological sections showed no noticeable degradation of alginate within the three-month observation period. Although there was an initial modest foreign body reaction, after this had subsided, the implants became embedded in thin fibrous sheaths, which were infiltrated with vascular channels and fibroblasts. The authors concluded that Kaltostat fibers in the rat model presented no obvious

Alginate Fibers and Wound Dressings: Seaweed Derived Natural Therapy, First Edition. Yimin Qin.

toxic risk or contraindication to their use as wound dressings or as hemostatic agents in general surgery.

A novel freeze-dried alginate gel dressing (AGA-100) was compared with extracts prepared from Kaltostat and the latter was found to induce cytopathic effects when tested in vitro on L929 cells (mouse fibroblasts) [72]. In a second in vivo study, samples of both alginates together with cotton gauze were applied to circular full-thickness wounds on the backs of pigs. Wound tissue was harvested on day 18 for histological examination. The wounds dressed with AGA-100 showed rapid wound closure compared with the control wounds, dressed with Kaltostat and cotton gauze. Foreign-body reaction was marked in Kaltostat and gauze-treated wounds, but not in the wounds dressed with AGA-100. Based on these data, the authors conclude that the use of AGA-100 could reduce cytotoxicity to fibroblasts and foreign body reactions that have been observed with currently available calcium alginate.

It has been observed that alginate-based microcapsules containing islets of Langerhans used as a bioartificial pancreas produce a foreign body reaction with fibrosis in an animal model. Pueyo et al. [56] demonstrated that macrophage cells involved in this process could be produced from monocytes activated by alginate-polylysine microcapsules in vitro. Otterlei et al. [53] compared the ability of alginates to stimulate human monocytes to produce three important cytokines, i.e. tumor necrosis factor-α (TNF-α), interleukin-1 and interleukin-6. They reported that high-M alginates (high in mannuronic acid) were approximately 10 times more potent in inducing cytokine production than high-G alginates (high in guluronic acid) and therefore proposed that mannuronic acid residues are the active cytokine inducers in alginates.

Other authors have also produced evidence to suggest that it is the ß(1–4) glycosidic linkage (M blocks) rather than the α(1–4) linkage (G blocks) that is responsible for cytokine stimulation and anti-tumor activity. These ß(1–4) bonds are found linking D-glucuronic acid in C-6-oxidized cellulose, which also has demonstrable TNF-α-stimulating activity, although this is limited compared with that of alginate rich in mannuronic acid [54]. Skjak-Braek and Espevik [67] reported that β(1–4)-linked uronic-acid polymers such as poly M are potent cytokine inducers in vivo, able to protect mice against lethal infections with *Staphylococcus aureus* or *Escherichia coli*. It is also stated that they can provide a marked degree of protection against lethal irradiation by increasing the production of myeloid blood cells as a result of stimulating hematopoietic cells in the bone marrow. Further evidence for the importance of high concentrations of M blocks comes from the finding that treatment of alginate with a high mannuronic acid content with C-5 epimerase, which converts β-D-mannuronic acid into α-L-guluronic acid, results in a loss of TNF-inducing ability. It is interesting to note that as early as in 1949, Rumble [59] recognized the importance of selecting the right type of alginate for treating hemorrhage following tooth extraction to ensure rapid absorption of the fiber.

Zimmerman et al. [91] and Klock et al. [40] disputed the difference in activity of M and G alginates following studies in which they tested different types of alginates for mitogenic activity both in vivo and in vitro before and after purification by free-flow electrophoresis and dialysis. They found that material treated in this way lost all its mitogenic properties regardless of the M/G ratio of the raw material, and suggested that this activity could be partly due to oligomers of mannuronic or guluronic acids. They also identified positively charged fractions with strong mitogenic activity that they proposed was related to lipopolysaccharide (LPS) molecules, but this observation is not in accord with the earlier work by Otterlei et al. [54], who demonstrated that mitogenic activity in alginates was not inhibited by the addition of Polymyxin B, which they showed was able to inhibit LPS-induced cytokine production.

LPS consists of a lipid A, a core oligosaccharide, and a polysaccharide part of varying size and complexity. In their review, Skjak-Braek and Espevik [67] state that LPS and poly M alginate share a common binding site on the macrophage, reacting with the membrane protein CD-14, which is believed to have a broad specificity for compounds rich in various types of sugar residues. They also report that the binding of poly M and LPS to monocytes can be inhibited by addition of G-blocks. Unlike LPS, which can stimulate cells that do not express membrane CD-14, poly M is unable to stimulate cell types that lack this membrane protein. Skjak-Braek and Espevik [67] suggest that poly M could activate the nonspecific immune system, thus increasing protection against various types of infections.

The effect of calcium alginate dressings on other cell types was investigated by Doyle et al. [26], who showed that low concentrations of an extract of one alginate dressing (Sorbsan) stimulated human fibroblasts on extended contact but decreased the proliferation of human microvascular endothelial cells and keratinocytes. They proposed that this activity could be due to calcium ions released from the dressing during the gelling process.

8.3 Wound Healing Mechanisms of Alginate Wound Dressings

Wound dressings are used to facilitate the wound healing process, and several authors have identified their ideal performance characteristics [73, 82]. For alginate wound dressings, the primary wound care mechanism is the formation of a hydrophilic gel secondary to ionic exchange. Jeter and Tintle [38] discussed the actions and properties of the in situ formed alginate gel where the main characteristics include:

1. The gelling process wicks wound exudate into the dressing directly over the wound;
2. The gel maintains a physiologically moist environment of the wound surface that is beneficial to healing and granulation tissue formation;
3. The gel provides a non-adherent interface with the wound bed;
4. The gel allows for the exchange of gases to and from the wound surface;
5. The gel offers a physical barrier against inadvertent contamination;
6. The gel permits nontraumatic dressing changes because it can be rinsed away with saline solution.

Alginate dressing can absorb a high volume of exudate directly over the wound bed without wicking wound exudate horizontally onto intact skin, even when the dressing extended beyond the wound. The process by which the fibers of the dressing are manufactured and the nature of the absorbent material are such that it created a wall of gel around the wound edge that does not permit wound exudate to move laterally. McMullen [46] observed that no maceration of wound edges or of the skin adjacent to the wound occurred with alginate dressings since the formation of gel stopped at the wound edges. Fraser and Gilchrist [31] found that alginate fibers can be used as a most effective wick to drain and dispose of excess moisture without causing wound maceration.

Barnett and Varley [3], Dealey [24], and Chaloner [16] emphasized the effectiveness of alginate wound dressings in healing partial thickness and cavity wounds. Favorable clinical results were reported for both types of wounds with exudate removal cited as a major factor in wound healing. Saline was effective to dissolve alginate dressings and dressing change is achieved painlessly and without disturbing the new layer of growing cells. The adherent dressings are easily removed by irrigation with sterile normal saline, which converts the insoluble calcium salt to the soluble sodium alginate. Fibers trapped in the wound do not have to be removed as they are easily dissolved by saline during dressing changes, thus avoiding disturbance of healing granulation tissue. In some instances, the alginate dressings play a part in controlling microbiological contamination of the wound by a dual process of physical entrapment and by exerting an antimicrobial effect by virtue of the presence of lysozyme from the exudate, which may be incorporated into the gel.

The single most frequently mentioned factor involving alginate wound dressings in wound care was reduced pain during application and removal. McMullen [46] reported that patients who had previously experienced pain with their dressing changes prior to using the calcium alginate dressing stated that their discomfort was dramatically less, especially those patients with lower extremity ulcerations.

Alginate wound dressings are also known to be able to accelerate wound healing. Thomas and Tucker [80] found that among 64 patients, the healing rate for alginate dressing (Sorbsan) treated patients was significantly greater than the paraffin gauze group where the percentage of wounds, which respond positively to treatment with Sorbsan is almost twice that obtained with the paraffin tulle, and the healing rates recorded for the Sorbsan treated group are almost five times greater.

8.4 Clinical Applications of Alginate Wound Dressings

Alginate wound dressings are now widely used around the world and there have been many publications on their clinical applications [18, 22, 27, 32, 36, 45–47, 49, 50, 63, 76, 89]. These novel dressings have been successfully used in the management of pressure sores, leg ulcers, diabetic foot ulcers, burn wounds, donor sites, bleeding wounds, surgical wounds, and many other types of superficial and cavity wounds. Figure 8.1 illustrates the use of alginate flat dressing and cavity filler.

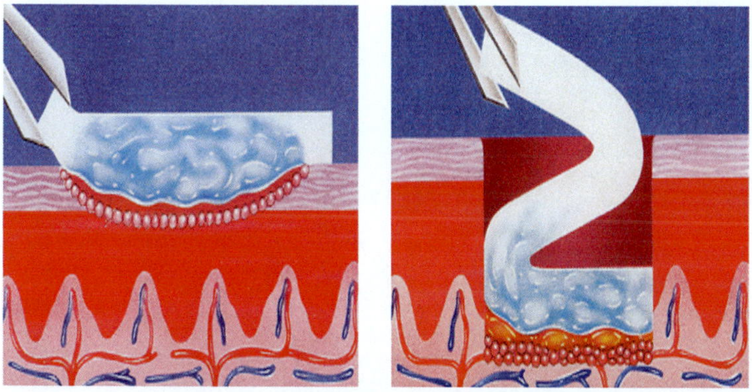

Figure 8.1 An illustration of the applications of alginate (a) flat dressing and (b) cavity filler.

8.4.1 Applications of Alginate Wound Dressings in Pressure Ulcers

Sayag et al. [60] carried out a prospective, randomized, controlled trial of 92 patients with full-thickness pressure ulcers and compared the efficacy of an alginate wound dressing with an established local treatment with dextranomer paste. During the treatment, a minimal 40% reduction in wound area was obtained in 74% of the patients in the alginate group and in 42% of those in the dextranomer group. The median time taken to achieve this goal was four weeks with alginate and more than eight weeks in the control group. Mean surface area reduction per week was 2.39 and 0.27 cm^2 in the alginate and dextranomer groups respectively. This difference was highly significant when the subgroups of almost completely healed subjects at the end of the study were considered. This striking healing efficacy of the alginate dressing suggests it possesses some pharmacological properties.

Chapuis and Dollfus [17] used calcium alginate wound dressing (Sorbsan) to treat 19 patients with 30 pressure sores and concluded that the dressing appears to give good results in patients with spinal cord lesions, including odor control. Similar claims for a calcium/sodium alginate dressing (Kaltostat) in the management of pressure sores were made by other authors [29, 46, 50, 90].

8.4.2 Applications of Alginate Wound Dressings in Leg Ulcers

Thomas and Tucker [80] reported the first controlled trial of an alginate dressing involving 64 community patients with leg ulcers. The patients were allocated to treatment with either an alginate (Sorbsan) or paraffin gauze (tulle) as a control. Only 4% of the ulcers treated with tulle healed during the study, while 31% of those treated with the alginate healed completely. The average healing rate achieved with the alginate was over four times that with tulle. Overall, 73% of alginate patients showed improvement during the trial, compared with 43% in the control group. However, this study was criticized on a number of grounds, principally because patients were not given sustained graduated compression [30, 69]. At the time, however, no competent compression bandages were available on the British Drug Tariff, and for this

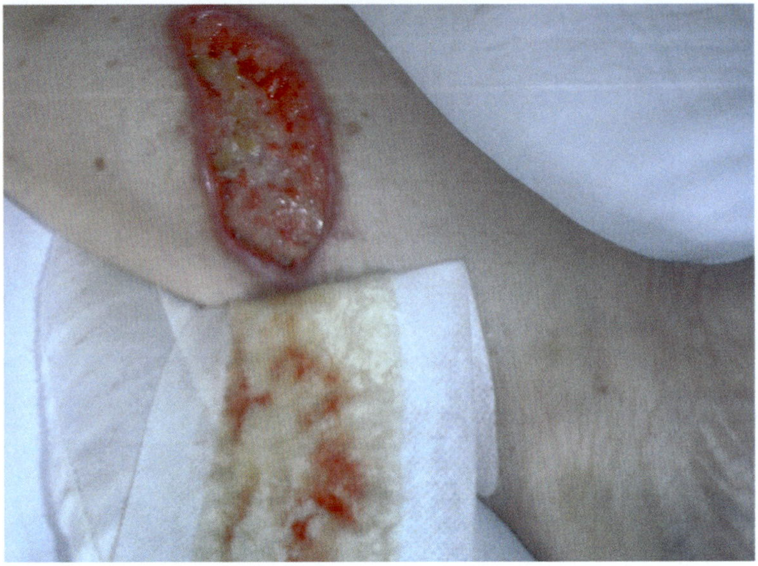

Figure 8.2 A leg ulcer wound covered by a piece of alginate wound dressing.

reason, they were excluded from the protocol [74]. Figure 8.2 shows a leg ulcer wound covered by a piece of alginate wound dressing.

Moffatt et al. [48] performed a randomized controlled trial comparing an alginate with a simple knitted viscose (NA) primary dressing under graduated compression bandaging. Sixty patients were randomized to treatment, where 26 dressed with the alginate healed, compared with 24 dressed with NA. This difference was however not statistically significant.

Scurr et al. [61] compared Sorbsan with a hydrocolloid (Granuflex) in 40 patients with venous ulcers. Compression was provided by a Class III compression stocking. Patients were evaluated weekly for six weeks or until healed. Of the wounds dressed with the alginate, six healed and 70% improved (decreased in size by more than 40%), while for those dressed with hydrocolloid, two healed and 45% improved. These differences were however not statistically significant. Pain scores were significantly lower for patients dressed with the alginate and all but two of those managed with the hydrocolloid had maceration around the wound.

Stacey et al. [70] compared Kaltostat, a zinc paste bandage (Viscopaste), and a zinc oxide-impregnated stockinette in a randomized controlled trial involving 113 patients with 133 ulcerated limbs. A minimal-stretch bandage (Elastocrêpe) provided compression with an elasticated tubular stockinette (Tubigrip) over the top to hold the dressings in place. Patients were followed for nine months or until the limbs had healed. Only ulcer size, ulceration on the right leg, and use of the paste bandage had any significant effect on the time to healing. At the 12-week stage, 64% of ulcers dressed with the paste bandage had healed, compared with 50% of those dressed with the alginate and the zinc oxide stockinette. The authors proposed that the difference in healing could be due to the extra layers of bandage that result from the use of the paste bandage, and suggested that these may have produced sustained levels of compression higher than the other two systems.

Armstrong and Ruckley [1] compared Sorbsan with a fibrous dressing made from carboxymethylcellulose (Aquacel) in a randomized trial of 44 patients with exuding leg ulcers. No difference was detected between the products in terms of healing rates achieved, but a statistically significant difference was observed in mean wear time (four days in the hydrofiber group compared to three for the alginate).

8.4.3 Applications of Alginate Wound Dressings in Diabetic Foot Ulcers

Neuropathy and peripheral vascular disease are common foot problems affecting 15–20% of diabetic patients. Following early success reported by Fraser and Gilchrist [31], alginates have been widely used for these and other foot wounds [11]. The low-adherent properties of alginates make them useful for dressing toes following surgery for ingrowing toenails. Burrow and Lindsay [12] compared Kaltostat with Ultraplast, an adhesive island dressing with an integral alginate pad. No major differences were detected in terms of healing rates, but Ultraplast caused more maceration of the toe and surrounding tissue. It was suggested that this was due to the adhesive carrier film. In other studies, alginates have reduced healing time and number of follow-up visits required [68], and healing time and number of dressing changes [28], compared to a perforated film dressing in patients undergoing partial or complete toenail avulsion.

A collagen–alginate dressing (Fibracol) was found to be superior to soaks and daily dressing changes in the postoperative management of chemical matricectomies, reducing the average healing time from 36 to 24 days [83]. Donaghue et al. [25] compared Fibracol with gauze moistened with normal saline for the treatment of diabetic foot ulcers. Seventy-five patients were assigned randomly in a 2 : 1 ratio to receive the alginate or gauze. Participants were seen weekly for a maximum of eight weeks or until the wounds healed. Although the mean reduction in wound area and incidence of complete healing were higher in the alginate group, the differences did not reach statistical significance. At the end of the study, mean reduction of wound area was $80.6 \pm 6\%$ in the alginate group and $61.1 \pm 26\%$ in the gauze group. In the alginate group, 48% (24/50) of patients had complete ulcer healing, compared to 36% (9/25) in the gauze group, although the mean time to healing was longer for the alginate group (6.2 ± 0.4 weeks vs 5.8 ± 0.4 weeks).

8.4.4 Applications of Alginate Wound Dressings in Burn Wounds and Donor Sites

The hemostatic, absorbent, and low-adherent properties of alginate wound dressings are ideally suited for the treatment of burns and donor sites. Groves and Lawrence [33] compared Sorbsan with a standard gauze pad. In a laboratory test, the alginate dressing absorbed nearly three times as much citrated blood as the gauze. When applied to fresh, split-thickness donor sites for five minutes after excision, blood loss from the sites treated with the alginate was almost half that from those treated with gauze.

The effect of longer-term application of alginate dressings to donor sites was investigated by Attwood [2]. In an initial study, 15 patients with split skin grafts had half of their wound dressed with an alginate (Kaltostat), and the other with paraffin gauze. Every area dressed with the alginate showed significantly better healing than the corresponding "control" area. The second phase of the study assessed the time to complete healing of alginate-dressed areas and patient acceptability of the dressing. A total of 155 donor sites were examined, where 130 were treated with alginate and 25 with paraffin gauze. Sites treated with the alginate healed in 7.0 ± 0.71 days, while paraffin gauze-dressed wounds took 10.75 ± 1.6. Patient comfort and quality of healing with the alginate were significantly better than that achieved with gauze. Attwood also discussed secondary dressings that could be used with Kaltostat and suggested that it was not necessary to use the bulky dressings commonly used when dressing donor sites. For some wounds, a film dressing applied over the alginate prevented desiccation and improved patient comfort, providing that hemostasis and good adhesion to the surrounding skin could be achieved. Wounds on the torso dressed with alginate were sometimes left uncovered.

Basse et al. [4] also compared Kaltostat with paraffin gauze in 17 patients with mirror-image donor sites on the thighs. Saline-soaked gauze was first applied to achieve hemostasis, followed by the appropriate dressing, covered by an elastic bandage. Because of the small numbers involved, no statistical tests were performed, but there was evidence to suggest that both blood loss and discomfort were reduced with the alginate. The mean time to healing for the alginate was 8.3 days (range 7–11), compared to 10.2 days for gauze (range 7–17).

O'Donoghue et al. [51] undertook a prospective randomized control trial involving 51 patients with donor sites, 30 of whom were randomized to treatment with calcium alginate and 21 to paraffin gauze. In one group, a single layer of paraffin gauze was applied, covered with three layers of cotton gauze and cotton wool padding, secured with a bandage. In the other group, Kaltostat impregnated with 0.25% bupivacaine, was overlaid with paraffin gauze, then with cotton gauze and padding as for the controls. Ten days post-harvesting, 21 of the 30 patients dressed with the alginate were completely healed, while only 7 of the 21 in the gauze group were healed. The authors considered that the choice of day 10 as the inspection period was probably inappropriate, following further investigations in which they removed the alginate on day 7 as recommended by Attwood. At this stage, the dressing was still moist and separated painlessly from the wound. By day 10 it had dried out and was more difficult to remove.

Cihantimur et al. [19] carried out a prospective, randomized, and controlled study to compare Kaltostat with paraffin gauze in the treatment of split-thickness skin graft donor sites in 40 patients. The mean time from operation to observation of complete healing was 8.5 days with the alginate and 11.5 days with gauze. Patient comfort and the quality of regenerated skin were better in the sites dressed with the alginate.

Rives et al. [57] compared the use of calcium alginate dressing and paraffin gauze in the treatment of scalp donor sites in 67 children in a controlled, randomized, clinical trial. Epithelialization occurred after 10 days in the alginate group and 11 in the gauze group (not significant). Earlier reharvesting of the donor site was possible in the alginate group than in the control group, and removal of the alginate dressing caused significantly less trauma and pain than the gauze.

Steenfos and Agren [71] compared a fiber-free alginate dressing (Comfeel SeaSorb) with paraffin gauze on donor sites in 17 patients. Both dressings were applied to parts of each wound and covered with gauze and a crêpe bandage. The alginate absorbed 40% more blood during the first 10 minutes post-wounding than the gauze. Examination of punch biopsies from 10 wounds on day 6 showed that 9 of the areas treated with the alginate had fully epithelialized compared with 7 treated with gauze. This difference did not reach statistical significance. The authors concluded that although the alginate showed increased initial blood absorption and quicker hemostasis, it had no clear beneficial effect on epithelialization. They suggested that this might have been due to the small sample size and to their failure to occlude the alginate over the critical first two postoperative days.

Calcium alginate was compared with Scarlet Red ointment dressing by Bettinger et al. [7] in 12 paired wounds in 7 patients undergoing skin grafting. No significant differences in healing time were recorded, with both groups taking 11.8 days. However, the dressing procedures used may have extended the healing time of the alginate. The same two dressings were compared in a second study by Lawrence and Balke [43] in which 46 patients had split-thickness skin grafts harvested from the upper inner thigh. Kaltostat and Scarlet Red were each applied to half of the wound and covered with gauze, cotton wool, and a crêpe bandage. The dressings were changed after 10 days when healing was assessed. The ointment was significantly better than the alginate, with 84% and 72% having healed respectively.

The relative merits of Kaltostat and a porcine xenograft (E-Z Derm) were examined in a controlled, prospective study of split-thickness skin graft donor sites on 20 patients [84]. Time to complete healing, quality of regenerated skin, and patient comfort were assessed. Time to healing was 8.1 days with the alginate and 11.3 with the xenograft. The quality of the healed skin under the alginate was superior to that under the xenograft in 95% of patients. No hypertrophic scarring was noted in patients treated with alginate, compared to 25% of xenograft sites.

Kelly et al. [39] used the gel-forming properties of alginates to protect exposed tissue, using them as a temporary recipient bed dressing prior to the delayed application of split skin grafts.

Deep burns of the hand represent a common serious surgical problem with major occupational and economic implications. Alginate dressings help to control hemorrhage during excision and grafting and prevent desiccation of important deep structures such as extensor tendons or joints that may be exposed during surgery [41].

A novel use of an alginate in the management of extensive burns on a seven-year-old boy was described by Varma et al. [85]. In the absence of sufficient donor site material to cover the area with a conventional graft, small skin grafts (edges of meshed grafts and shredded skin) were cut into small pieces, formed into a suspension, and poured evenly over a sheet of Kaltostat, which was applied to the wound. Epithelial islands were visible after 10 days and complete coverage was achieved by day 28.

The problem of donor-site pain was addressed by Butler et al. [13] in a study involving 45 patients undergoing split-thickness skin grafting. After harvesting the graft, each patient was randomized to dry Kaltostat, Kaltostat moistened with 20 ml of saline, or Kaltostat moistened with 20 ml of bupivacaine 0.5%. The dressing was covered with an outer wound pad and secured with an adhesive dressing. A blinded medical observer assessed postoperative pain at 24, 48, and 72 hours using a linear analog scale. The dressing was removed on day 10 and healing was assessed, when ease of removal and the presence of infection were also recorded. There was a dramatic statistical difference in pain scores between patients who received the bupivacaine-soaked dressing and the other two groups at 24 and 48 hours. However, by 72 hours all patients reported only low levels of discomfort. No differences in ease of removal were recorded.

Porter [55] compared alginate dressings with hydrocolloid dressings in the healing of split skin graft donor sites. Sixty-five patients were randomized to treatment and the rates of epithelialization, patient discomfort, and convenience of clinical use were compared. The alginate dressing was applied to the raw donor areas and held in place by layers of dry gauze, plaster wool, and a crêpe bandage. The hydrocolloid was applied after hemostasis was achieved, and covered by plaster wool and a crêpe bandage. At the first dressing change, 87% of the donor areas dressed with the hydrocolloid, and 86% of those dressed with the alginate were more than 90% healed. The mean time from operation to complete healing was 10 days for areas dressed with the hydrocolloid and 15.5 days for those dressed with alginate. This difference was statistically significant. It is considered that the poor performance of the alginate dressing was due to dressing technique.

8.4.5 Applications of Alginate Wound Dressings as a Hemostatic Agent for Bleeding Wounds

There is a long history of alginate fibers being used as a hemostatic agent. In 1951, Blaine performed a comparative evaluation of absorbable hemostatic agents that included alginates [8]. Although calcium alginate fibers took up to 12 weeks to be fully absorbed, sodium calcium alginates were generally absorbed within 10 days. In experiments where there was minimal trauma involving small implants, uneventful absorption took place within a few days. No evidence of adverse local histological changes were detected, although the rate of absorption varied with the location and vascularity of the surrounding tissue. A further study showed that in the presence of antiseptic agents such as cetylpyridinium bromide, the absorption process was generally incomplete, and some histological changes were noted including encapsulation and giant cell production [52].

In a more recent study, Segal et al. [62] showed that alginate materials activated coagulation more than non-alginate materials. The extent of coagulation activation was affected differently by the alginate M or G group composition. It was demonstrated that alginates containing zinc ions had the greatest potentiating effect on prothrombotic coagulation and platelet activation.

As a hemostatic agent, the lack of adverse tissue reactions to alginates was described by Jaros and Dewey [37], following long-term administration of sodium alginate as an adjuvant for repository hyposensitization agents. The material was well tolerated with less of a reaction rate than expected using regular allergenic extracts.

Blair et al. [9] compared the hemostatic effect of four different materials in liver lacerations in the rabbit. The products examined were oxidized cellulose (Surgicel), porcine collagen (Medistat), calcium alginate (Kaltostat), and surgical gauze. Calcium alginate stopped bleeding in less than three minutes compared with a mean of 5.7 ± 0.75 minutes for porcine collagen, 12.5 ± 0.9 minutes for oxidized cellulose, and >15 minutes with gauze. Oxidized cellulose and calcium alginate reabsorbed within three months, leaving a fibrous scar, but a vigorous foreign body reaction was observed with porcine collagen. In a further group of animal samples, all three hemostatic agents were left in situ on the liver and small bowel mesentery for three months. The porcine collagen caused fatal intestinal obstruction in five animals.

Sirimanna [64] investigated the use of calcium alginate fiber for packing nasal cavity following surgical trimming of the inferior turbinates. Thirty-two nostrils were packed with Kaltostat for 36–48 hours to achieve hemostasis. There was no bleeding while the packs were in place or after removal. These results were compared retrospectively with two other treatments, i.e. trousered paraffin gauze and glove finger packs, both of which had been associated with bleeding in over 50% of cases either while in situ or after removal.

In a second study, the three types of packing were compared prospectively [65]. All three were similarly effective in preventing bleeding while in situ, but the alginate caused significantly less bleeding on removal. Reduced bleeding was recorded with all three packs if left for 48 hours. It was suggested that the observed reduction in bleeding might lead to a decrease in postoperative infections. In a follow-up study on the same group of patients over a three-week period [66], packing for 48 hours resulted in significantly more complications for all three materials than when packs were removed at 24 hours. It appeared that more infection, crusting, and airway problems were associated with use of the alginate than the other two treatments if the packs were left in place for 48 hours. However, patient numbers were too small to test this statistically.

The exchange of calcium ions with sodium ions in the blood is commonly known as the main reason for hemostasis efficacy of alginate wound dressings and it is reasonable to suppose that products made from alginates containing a higher proportion of mannuronic acid residues, which give up their calcium ions more readily than those high in guluronic acid, could be more effective hemostats. Evidence for the hemostatic activity of one high mannuronic alginate dressing, Sorbsan, was produced in a study in which the dressing was applied to experimental, full- and partial-thickness wound models for up to 14 days to assess its effects on healing. Histological evaluation showed the dressing to be an effective hemostat, generally well-tolerated by body tissues. Good epidermal healing was seen on all wounds, although cellular reactions could be provoked in full-thickness wounds without occlusion if there was an insufficient volume of exudate to completely wet the alginate fibers [3].

8.4.6 Applications of Alginate Wound Dressings in Surgical Wounds

Gupta et al. [35] studied the use of alginate fiber (Sorbsan) for packing cavities following surgery in 29 patients. Compared with proflavine-soaked gauze, the alginate dressing caused less pain and reduced the need for analgesia. Alginate dressing also appeared to reduce bacterial counts within the wound. The authors concluded that the advantages offered by the alginate dressing outweighed the increased cost.

In another study [23], 16 patients were randomized to receive calcium alginate dressing and 18 to saline-soaked gauze for packing abscess cavities following incision and drainage. At the first dressing change, the patient marked on a linear analog scale the pain experienced, while the nurse noted the ease of removal. Patients evaluated the alginate dressing as significantly less painful, and nurses found it easier to remove than the gauze.

Similar advantages for the use of a calcium alginate dressing (Sorbsan) over conventional paraffin gauze/cotton packs were reported following a prospective randomized study following hemorrhoidectomy [23]. Fifty patients were prospectively randomized to receive paraffin gauze/cotton gauze roll or alginate roll as a postoperative pack. Pain was assessed at six hours postoperatively, on removal of rectal packing and at first bowel action. Hemorrhage was monitored at six hours and on removal of pack. Patients who received the alginate dressing reported reduced pain at the time of removal/spontaneous discharge of rectal packing and first postoperative bowel action. No significant difference in postoperative hemorrhage, or hospital stay was reported.

Cannavo et al. [15] compared the performance of three different dressings in the management of 36 dehisced surgical abdominal wounds. These were a standard alginate dressing, a gauze moistened with sodium hypochlorite (0.05%), and a combined dressing pad, with the latter consisting of an absorbent pad to which is added a semi-permeable film dressing. No statistically significant differences in healing rates between the three treatment groups were detected but there was a trend for the combined dressing pad protocol to produce a greater reduction in wound area. Maximum pain was significantly greater and satisfaction significantly lower among patients who received the sodium hypochlorite treatment. The associated treatment costs were also substantially higher for this group of patients. The authors concluded that the use of sodium hypochlorite-soaked dressings for surgical wounds should be abandoned.

Patients with gaping abdominal wounds following cesarean section and radical vulvectomy [58] were also managed successfully with alginate dressings. A patient with a 10-year history of heroin abuse and multiple ulcerations to his upper arm had his wounds dressed with a calcium alginate rope and covered with a

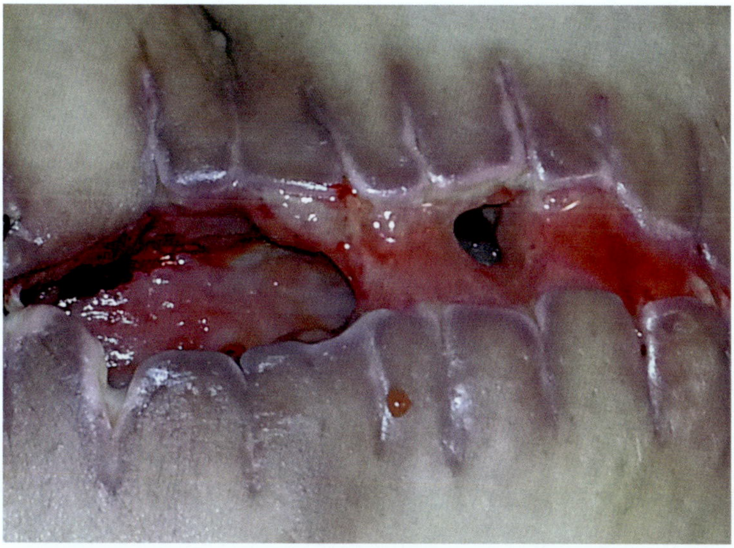

Figure 8.3 An illustration of fat liquefaction of incision wound.

four-layer bandage. Dressings were changed weekly and during treatment the patient remained heroin free. Complete healing was achieved in 42 days [20].

Fat liquefaction of incision wounds such as the one shown in Figure 8.3 commonly occurs after surgical operation, especially among overweight people. The high absorption capacity and gel-forming properties of alginate wound dressings are particularly suited for this type of wound. Gu et al. [34] compared the applications of alginate dressing and vaseline gauze on 31 cases of fat liquefaction incision after abdominal surgery. Results showed that after four weeks, the overall result of the alginate dressing group is clearly better than the control group of vaseline gauze. With alginate wound dressing, there was no adherence between dressing and wound bed, and the release of calcium ions improved blood clotting. On the other hand, the use of vaseline gauze caused pain and wound bleeding. As shown in Tables 8.1 and 8.2, the clinical efficacy of alginate wound dressing is advantageous to the traditional vaseline gauze.

Table 8.3 compares the clinical efficacy of the two test groups, showing that alginate dressings are far better than vaseline gauze in the treatment of incision wounds with fat liquefaction.

Table 8.1 A comparison of pain experienced during the treatment of fat liquefaction incision wound.

Group	Number of cases	No pain	Slight painful	Painful	Heavily painful
Alginate dressing	31	5	19	7	0
Vaseline gauze	30	0	6	19	5

Table 8.2 A comparison of the frequency of dressing change during the treatment of fat liquefaction incision wound.

Group	Number of dressing changes (first week)	Number of dressing changes (after one week)
Alginate dressing	Every three days	Every four days
Vaseline gauze	Daily	Every two days

Table 8.3 A comparison of clinical efficacy during the treatment of fat liquefaction incision wound.

	Cases showing different levels of clinical efficacy				
Group	Overall cases	Healed	Recovered	Improved	No effect
Alginate dressing	31	29	2	0	0
Vaseline gauze	30	9	6	8	7

8.4.7 Applications of Alginate Wound Dressings in Nose Surgery

Xie et al. [87] compared the efficacy of four commonly used nasal packing materials, i.e. vaseline gauze strips, rapid rhino nasal pac with gel knit, calcium alginate dressings, and expansive sponge. Table 8.4 shows a comparison of the degree of bleeding following removal of nasal packing materials. For the traditional vaseline gauze, not only 35 of the 40 patients under treatment felt pain on dressing removal, 38 of them continued to bleed upon dressing change. Calcium alginate dressing showed one of the best performances, with none of the 23 patients showing bleeding after removal of the dressing, and bleeding completely stopped among 22 of the 23 patients under treatment.

Liu and Zheng [44] used calcium alginate wound dressing to fill in one side and aureomycin gauze in another side during nasal surgery, and compared nasal

Table 8.4 A comparison of the degree of bleeding following removal of nasal packing materials.

	Degree of bleeding			
Nasal packing materials	No	Little	Yes	Overall cases
Vaseline gauze strips	2	24	14	40
Rapid rhino nasal pac with gel knit	32	3	0	35
Expansive sponge	19	13	6	38
Calcium alginate dressing	22	1	0	23

Table 8.5 A comparison of nasal mucosal edema and submucosal inflammatory cell infiltration.

		Degree of nasal mucosa edema			Degree of inflammatory cell infiltration		
Group	**Cases**	**Light**	**Medium**	**Heavy**	**Light**	**Medium**	**Heavy**
Calcium alginate dressing	53	19	28	6	16	30	7
Aureomycin gauze	53	10	29	14	15	26	12

mucosal edema and submucosal inflammatory cell infiltration. Results showed that during removal of the dressing, the level of pain was much lower with the alginate dressing. As shown in Table 8.5, the overall efficacy is much better for the alginate dressing than aureomycin gauze.

8.4.8 Applications of Alginate Wound Dressings in Anal Fistula Surgery

Wei et al. [86] compared alginate wound dressing and traditional gauze dressing on mixed hemorrhoid, anal fistula and prolapsed hemorrhoid wounds after surgery. One hundred and sixty cases were split into two groups with 90 cases treated with alginate dressing and the other 70 treated with iodoform gauze. Tables 8.6–8.8 compared the pain during application and removal of the dressing, bleeding, and patient comfort during treatment. Results showed that the overall clinical efficacy of alginate dressing is much better than the conventional iodoform gauze.

In anal fistula surgery, the filling of surgical wounds with alginate wound dressing can relieve pain and improve the process of application and removal of the cavity-filling process. While conventional method of filling the surgical

Table 8.6 A comparison of pain during treatment.

Group	Total number of cases	No pain	Slight pain	Medium level of pain	Highly painful
Alginate dressing	90	41	31	15	3
Iodoform gauze	70	6	12	46	6

Table 8.7 A comparison of bleeding during treatment.

Group	Total number of cases	No bleeding (%)	Slightly bleeding (%)	Bleeding (%)
Alginate dressing	90	81.1	16.7	2.2
Iodoform gauze	70	24.3	58.6	17.1

Table 8.8 A comparison of patient comfort during treatment.

Group	Total number of cases	Bulge feeling (%)	Odor (%)	Uroschesis (%)	Tamponade tolerance ≤12 h (%)
Alginate dressing	90	18.9	10.0	7.8	6.7
Iodoform gauze	70	87.1	85.7	25.7	68.6

wound with iodoform gauze can have antimicrobial and other clinical efficacy, its low absorbency and poor adherence properties can cause pain and discomfort to patients undergoing such operations.

8.4.9 Applications of Alginate Wound Dressings in Cavity Wounds

Berry et al. [5] compared Kaltostat with a polyurethane foam dressing (Allevyn) in the management of patients with non-infected cavity wounds. Both dressings were found to be easy to use, effective, and acceptable to patients and clinicians. A controlled trial set out to compare calcium alginate dressing with more traditional saline-soaked gauze for packing abscess cavities, following incision and drainage. Patients were randomized to receive either calcium alginate dressing (16 patients) or gauze dressing (18 patients). At the first dressing change, the patient marked on a linear analog scale the pain experienced, while the nurse noted similarly the ease of removal of the dressing. Calcium alginate dressing was significantly less painful to remove after operation, and also easier to remove than gauze dressings. If abscess cavities are packed after incision and drainage, calcium alginate dressing appears to be an improvement on conventional dressings [23].

8.4.10 Applications of Alginate-Based Composite Wound Dressings

Composite dressings containing alginate have been developed in a variety of forms that include simple adhesive island dressings (Kaltoclude) to absorbent pads with an alginate wound contact layer (Sorbsan Plus). An alginate-faced dressing containing activated charcoal for use in the management of malodorous wounds has also been developed and the results of a laboratory-based evaluation to compare the performance of this dressing with other charcoal dressings has been described in the literature [81].

8.5 Main Properties of Alginate Wound Dressings

Calcium alginate fibers are chemically the salt between calcium ion and alginic acid and upon contact with wound exudate, the calcium ions in the fiber exchange with sodium ions in the exudate, resulting in the formation of sodium alginate which is water soluble and can absorb a large amount of water into the fiber structure, causing

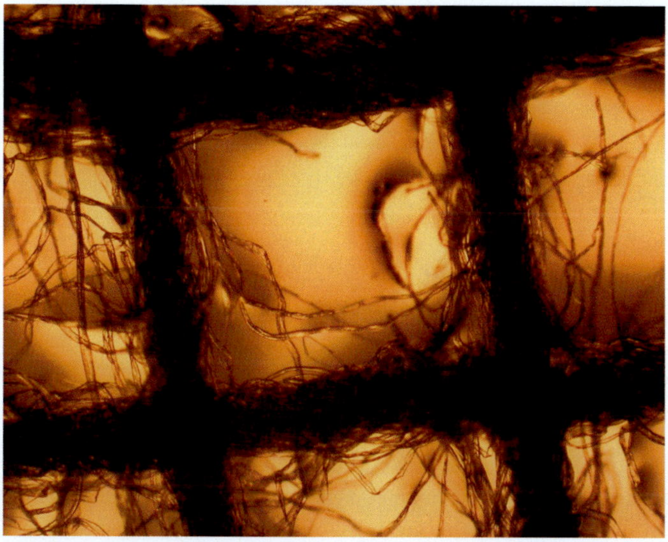

Figure 8.4 Cotton gauze wet in normal saline.

the swelling of the fiber and the in situ formation of a fibrous hydrogel on the wound surface. This unique gelling property gives alginate wound dressing a number of clinical benefits in wound management that outperforms traditional wound management materials such as cotton gauzes.

As shown in Figure 8.4, when wet in normal saline, the absorption of cotton gauze takes place in the capillary spaces between the yarns. Although absorption capacity can be substantial, the liquid held by the cotton fibers can easily spread along the textile structure, often from the exuding wound to surrounding areas, causing maceration to healthy skin around the wound. On the other hand, alginate fibers swell greatly when in contact with wound exudate, and subsequently, the lateral spreading of wound exudate is blocked because of the closure of the spaces between fibers, resulting in the "gel blocking" property of alginate wound dressings as can be seen in Figure 8.2.

After many years of clinical applications, alginate wound dressings have been proven to have a number of wound-healing properties, some of which are summarized below.

8.5.1 Wound-Healing Promotion

Berven et al. [6] showed that as a marine bioactive substance, alginate has chemotactic properties and can accelerate the wound healing process by attracting wound site cells that are involved in the wound repair process. Attwood [2] compared alginate dressing and conventional gauze dressing on donor site wounds. Results showed that among the 130 wounds, the time needed for complete healing reduced from 10 days for traditional gauze to 7 days for alginate wound dressings.

In another study, Sayag et al. [60] showed that under the same condition, 74% of the patients had their wound area reduced by 40% when treated with alginate dressing, while only 42% of the patient had this outcome when using tradition gauze. The application of alginate wound dressing can reduce the healing time by eight weeks.

Doyle et al. [26] studied the effect of alginate wound dressing on fibroblast cells, endothelial cells, and keratinocytes. Results showed that alginate fibers can promote the growth of fibroblast cells, while inhibit the growth of endothelial cells and keratinocytes. Overall, the application of alginate wound dressings can promote the wound-healing process.

8.5.2 The Hemostatic Properties of Alginate Wound Dressing

Alginate wound dressings possess excellent hemostatic properties due to a number of factors such as the rapid swelling of the fibers upon contact with blood and the release of calcium ions through ion-exchange. Groves and Lawrence [33] found that within five minutes of contact with bleeding wounds, the alginate wound dressings had affected blood clotting. Segal et al. [62] assessed the hemostatic properties of several types of alginate wound dressings with different chemical structures and found that the hemostatic properties of alginate wound dressings were due to two mechanisms, i.e. blood coagulation and platelet activation, which is promoted through the action of calcium ions released from the dressing. It was found that zinc alginate wound dressings had better hemostatic properties than calcium alginate, mainly because the zinc ions are better at blood coagulation and platelet activation.

Davies et al. [21] compared the hemostatic effect of alginate wound dressing and a traditional gauze and found that on average, the blood loss during surgery with gauze is 139.4 ± 9.6 ml, while it was 98.8 ± 9.9 ml with alginate dressing. After the surgical operation, the blood loss was 158.4 ± 17.3 ml with gauze and it was 96.6 ± 11.7 ml with alginate dressing, showing the excellent hemostatic property of alginate wound dressing over conventional wound management products.

8.5.3 Pain Relief Properties of Alginate Wound Dressing

Butler et al. [14] studied the efficacy of alginate wound dressings on skin donor sites. They found that when the alginate dressing is soaked in hypochlorite solution and applied to wound, there was an obvious relief of pain. A prospective double-blind controlled trial examined the differences in postoperative split skin graft donor site pain between sites dressed with three differently treated types of dressing, i.e. a dry calcium alginate dressing, a saline-moistened calcium alginate dressing, and a bupivacaine hydrochloride (0.5%) moistened calcium alginate dressing. There was a significant reduction in postoperative pain in the calcium alginate and bupivacaine group (group 3) at 24 and 48 hours when compared to the other two groups. There was no difference in ease of removal of dressings or the quality of wound healing on day 10 between the three groups. This study suggested a significant reduction in postoperative pain in bupivacaine-soaked calcium alginate, without reducing the beneficial effects of the calcium alginate on donor site healing. Bettinger et al. [7] found similar pain relief effect when applying alginate wound dressings on burn wounds.

8.5.4 The Antimicrobial Properties of Alginate Wound Dressing

Because of the swelling of alginate fibers upon contact with wound exudate, the capillary space between fibers are closed and the bacteria attached on the fibers are trapped, with their ability to grow and cause infection inhibited. Bowler et al. [10] placed alginate dressing in a test solution containing bacteria and found that the alginate dressing can suppress bacteria growth. Young [89] used alginate dressing on infected and highly exuding wounds and achieved good clinical results.

8.5.5 Alginate Wound Dressings as Cavity Filler

Barnett and Varley [3], Dealey [24], and Chaloner [16] applied alginate wound dressings on cavity wounds. They found that the high absorption and gelling properties of alginate wound dressings are particularly suited to the management of cavity wounds, whereby the dressings can effectively fill the cavity and are easy to remove. Xie et al. [88] used alginate dressing in nasal surgery and compared its clinical efficacy to three other commonly used materials. The alginate dressing relieved pain and reduced blood loss and generally outperformed Vaseline gauze and expansive foam.

8.5.6 Cost-Effectiveness of Alginate Wound Dressings

Although the individual piece of alginate dressing is more expensive than conventional gauze products, the application of alginate wound dressing is cost-efficient in that the overall cost of material and nursing time is low because of the reduced dressing change and fast wound healing [27, 50, 75, 77–79].

8.6 Summary

Because of the ion exchange between calcium ions in the fiber and sodium ions in the wound exudate, alginate fibers can form gel when applied to exuding wounds. The gelling process is accompanied by the absorption of wound exudate into the fiber structure and as the fibers swell, the capillary structure in the nonwoven wound dressing is closed, thereby blocking the lateral spreading of liquid. This unique "gel blocking" property of alginate wound dressings helps to reduce wound maceration. In addition, since alginate wound dressings have novel hemostatic and antimicrobial properties, especially with the addition of silver ions, they are ideally suited for the management of chronic wounds where a high level of exudate is common.

References

1 Armstrong, S.H. and Ruckley, C.V. (1997). Use of a fibrous dressing in exuding leg ulcers. *J. Wound Care* 6 (7): 322–324.

2 Attwood, A.I. (1989). Calcium alginate dressing accelerates split-skin graft donor site healing. *Br. J. Plast. Surg.* 42 (4): 373–379.
3 Barnett, S.E. and Varley, S.J. (1987). The effects of calcium alginate on wound healing. *Ann. R. Coll. Surg. Engl.* 69 (4): 153–155.
4 Basse, P., Siim, E., and Lohmann, M. (1992). Treatment of donor sites: calcium alginate versus paraffin gauze. *Acta Chir. Plast.* 34 (2): 92–98.
5 Berry, D.P., Bale, S., and Harding, K.G. (1996). Dressings for treating cavity wounds. *J. Wound Care* 5 (1): 10–17.
6 Berven, L., Solberg, R., Truong, H.T. et al. (2013). Alginates induce legumain activity in RAW 264.7 cells and accelerate autoactivation of prolegumain. *Bioact. Carbohydr. Dietary Fibre* 2: 30–44.
7 Bettinger, D., Gore, D., and Humphries, Y. (1995). Evaluation of calcium alginate for skin graft donor sites. *J. Burn Care Rehabil.* 16 (1): 59–61.
8 Blaine, G. (1951). A comparative evaluation of absorbable haemostatics. *Postgrad. Med. J.* 27: 613–620.
9 Blair, S.D., Backhouse, C.M., Harper, R. et al. (1988). Comparison of absorbable materials for surgical haemostasis. *Br. J. Surg.* 75 (10): 969–971.
10 Bowler, P.G., Jones, S.A., and Davies, B.J. (1999). Infection control properties of some wound dressings. *J. Wound Care* 8 (10): 499–502.
11 Bradshaw, T. (1989). The use of Kaltostat in the treatment of ulceration in the diabetic foot. *Chiropodist* 44 (9): 204–207.
12 Burrow, B.A. and Lindsay, A. (1989). A limited evaluation of alginates and a small scale comparison between Kaltostat and a standard non-adherent dressing, Ultraplast Alginate, in the treatment of nail avulsion by matrix phenolisation. *Chiropodist* 3 (10): 211–218.
13 Butler, P.E., Eadie, P.A., Lawlor, D. et al. (1993). Bupivacaine and Kaltostat reduces postoperative donor site pain. *Br. J. Plast. Surg.* 46 (6): 523–524.
14 Butler, P.E., Eadie, P.A., Lawlor, D. et al. (1993). Calcium alginate dressing accelerates split skin graft donor site. *Br. J. Plast. Surg.* 46 (6): 523–524.
15 Cannavo, M., Fairbrother, G., Owen, D. et al. (1998). A comparison of dressings in the management of surgical abdominal wounds. *J. Wound Care* 7 (2): 57–62.
16 Chaloner, D. (1991). Treating a cavity wound. *Nurs. Times* 87: 67–69.
17 Chapuis, A. and Dollfus, P. (1990). The use of a calcium alginate dressing in the management of decubitus ulcers in patients with spinal cord lesions. *Paraplegia* 28 (4): 269–271.
18 Choate, C.S. (1994). Wound dressings. A comparison of classes and their principles of use. *J. Am. Podiatr. Med. Assoc.* 84 (9): 463–469.
19 Cihantimur, B., Kahveci, R., and Ozcan, M. (1997). Comparing Kaltostat with Jelonet in the treatment of split-thickness skin graft donor sites. *Eur. J. Plast. Surg.* 20 (5): 260–263.
20 Cortimiglia-Bisch, L. and Brazinsky, B. (1998). Use of a four-layer bandage system in the treatment of an intravenous drug abuser with chronic upper extremity ulcerations: a case study. *Ostomy Wound Manage* 44 (3): 48–52. 54–45.
21 Davies, M.S., Flannery, M.C., and McCollum, C.N. (1997). Calcium alginate as haemostatic swabs in hip fracture surgery. *J. R. Coll. Surg. Edinb.* 42 (1): 31–32.

22 Davis, E. (1993). Care of a patient with rheumatoid arthritis and a vasculitic ulcer. *J. Wound Care* 2 (2): 72–73.
23 Dawson, C., Armstrong, M.W., Fulford, S.C. et al. (1992). Use of calcium alginate to pack abscess cavities: a controlled clinical trial. *J. R. Coll. Surg. Edinb.* 37 (3): 177–179.
24 Dealey, C. (1989). Management of cavity wounds. *Nursing* 3 (39): 25–27.
25 Donaghue, V.M., Chrzan, J.S., Rosenblum, B.I. et al. (1998). Evaluation of a collagen-alginate wound dressing in the management of diabetic foot ulcers. *Adv. Wound Care* 11 (3): 114–119.
26 Doyle, J.W., Roth, T.P., Smith, R.M. et al. (1996). Effects of calcium alginate on cellular wound healing processes modeled in vitro. *J. Biomed. Mater. Res.* 32 (4): 561–568.
27 Fanucci, D. and Seese, J. (1991). Multi-faceted use of calcium alginates. *Ostomy Wound Manage* 37: 16–22.
28 Foley, G.B. and Allen, J. (1994). Wound healing after toenail avulsion. *Foot* 4: 88–91.
29 Fowler, E. and Papen, J.C. (1991). Evaluation of an alginate dressing for pressure ulcers. *Decubitus* 4 (3): 47–52.
30 Franks, P.J. (1990). Sorbsan and leg ulcers (letter). *Pharm. J.* 244: 468.
31 Fraser, R. and Gilchrist, T. (1983). Sorbsan calcium alginate fiber dressings in footcare. *Biomaterials* 4 (3): 222–224.
32 Gensheimer, D. (1993). A review of calcium alginates. *Ostomy Wound Manag.* 39 (1): 34–38.
33 Groves, A.R. and Lawrence, J.C. (1986). Alginate dressing as a donor site haemostat. *Ann. R. Coll. Surg. Engl.* 68 (1): 27–28.
34 Gu, G., Niu, L., and Yan, J. (2015). Observation on the effect of dressing change with alginate dressing in fat liquefaction incision after abdominal surgery. *Women's Health Res.* (24): 214–215.
35 Gupta, R., Foster, M.E., and Miller, E. (1991). Calcium alginate in the management of acute surgical wounds and abscesses. *J. Tissue Viability* 1 (4): 115–116.
36 Imamura, Y., Fujiwara, S., Sato, T. et al. (1996). Successful treatment of toxic epidermal necrolysis with calcium sodium alginate fiber (letter). *Int. J. Dermatol.* 35 (11): 834–835.
37 Jaros, S.H. and Dewey, J.L. (1964). Use of an alginate in hyposensitization. *Ann. Allergy* 22: 173–179.
38 Jeter, K.F. and Tintle, T.E. (1990). Early experience with a calcium alginate dressing. *Ostomy Wound Manag.* 28: 75–81.
39 Kelly, S.A., Dickson, M.G., and Sharpe, D.T. (1988). Calcium alginate as a temporary recipient bed dressing prior to the delayed application of split skin grafts. *Br. J. Plast. Surg.* 41 (4): 445.
40 Klock, G., Frank, H., Houben, R. et al. (1994). Production of purified alginates suitable for use in immunoisolated transplantation. *Appl. Microbiol. Biotechnol.* 40 (5): 638–643.
41 Kneafsey, B., O'Shaughnessy, M., and Condon, K.C. (1996). The use of calcium alginate dressings in deep hand burns. *Burns* 22 (1): 40–43.

42 Lansdown, A.B. and Payne, M.J. (1994). An evaluation of the local reaction and biodegradation of calcium sodium alginate (Kaltostat) following subcutaneous implantation in the rat. *J. R. Coll. Surg. Edinb.* 39 (5): 284–288.

43 Lawrence, J.E. and Blake, G.B. (1991). A comparison of calcium alginate and Scarlet Red dressings in the healing of split thickness skin graft donor sites. *Br. J. Plast. Surg.* 44 (4): 247–249.

44 Liu, Y. and Zheng, C. (2003). Clinical observation of the effect of alginate dressing in postoperative nasal cavity tamponade. *Chin. Arch. Otolaryngol. Head Neck Surg.* 10 (4): 211–215.

45 Marghoob, A.A., Artman, N.N., and Siegel, D.M. (1997). Calcium alginate dressings with second intention healing of surgical wounds: our experience. *Wounds* 9 (2): 50–55.

46 McMullen, D. (1991). Clinical experience with a calcium alginate dressing. *Dermatol. Nurs.* 3 (4): 216–219.

47 Miller, L., Jones, V., and Bale, S. (1993). The use of alginate packing in the management of deep sinuses. *J. Wound Care* 2 (5): 262–263.

48 Moffatt, C.J., Oldroyd, M.I., and Franks, P.J. (1992). Assessing a calcium alginate dressing for venous ulcers of the leg. *J. Wound Care* 1 (4): 22–24.

49 Morgan, D. (1996). Alginate dressings. *J. Tissue Viability* 7 (1): 4–14.

50 Motta, G.J. (1989). Calcium alginate topical wound dressings: a new dimension in the cost-effective treatment for exudating dermal wounds and pressure sores. *Ostomy Wound Manage* 25: 52–56.

51 O'Donoghue, J.M., O'Sullivan, S.T., Beausang, E.S. et al. (1997). Calcium alginate dressings promote healing of split skin graft donor sites. *Acta Chir. Plast.* 39 (2): 53–55.

52 Oliver, L.C. and Blaine, G. (1950). Haemostasis with absorbable alginates in neurosurgical practice. *Br. J. Surg.* 37: 307–310.

53 Otterlei, M., Ostgaard, K., and Skjak-Braek, G. (1991). Induction of cytokine production from human monocytes stimulated with alginate. *J. Immunother.* 10: 286–291.

54 Otterlei, M., Sundan, A., Skjak-Braek, G. et al. (1993). Similar mechanisms of action of defined polysaccharides and lipopolysaccharides: characterization of binding and tumor necrosis factor alpha induction. *Infect. Immun.* 61 (5): 1917–1925.

55 Porter, J.M. (1991). A comparative investigation of re-epithelialization of split skin graft donor areas after application of hydrocolloid and alginate dressings. *Br. J. Plast. Surg.* 44 (5): 333–337.

56 Pueyo, M.E., Darquy, S., Capron, F. et al. (1993). *In vitro* activation of human macrophages by alginate-polylysine microcapsules. *J. Biomater. Sci. Polym. Ed.* 5 (3): 197–203.

57 Rives, J.M., Pannier, M., Castede, J.C. et al. (1997). Calcium alginate versus paraffin gauze in the treatment of scalp graft donor sites. *Wounds* 9 (6): 199–205.

58 Roberts, K.J., Rowland, C.M., and Benbow, M.E. (1992). Managing a patient's infected wound site after a radical vulvectomy. *J. Wound Care* 1 (4): 14–17.

59 Rumble, J.F.S. (1949). Twenty-five cases treated with absorbable alginate wool. *Br. Dent. J.* 86: 203–205.

60 Sayag, J., Meaume, S., and Bohbot, S. (1996). Healing properties of calcium alginate dressings. *J. Wound Care* 5 (8): 357–362.

61 Scurr, J.H., Wilson, L.A., and Coleridge-Smith, P.D. (1994). A comparison of calcium alginate and hydrocolloid dressings in the management of chronic venous ulcers. *Wounds* 6 (1): 1–8.

62 Segal, H.C., Hunt, B.J., and Gilding, D.K. (1998). The effects of alginate and non-alginate wound dressings on blood coagulation and platelet activation. *J. Biomater. Appl.* 12 (3): 249–257.

63 Seymour, J. (1997). Alginate dressings in wound care management. *Nurs. Times* 93 (44): 49–52.

64 Sirimanna, K.S. (1989). Calcium alginate fiber (Kaltostat 2) for nasal packing after trimming of turbinates: a pilot study. *J. Laryngol. Otol.* 103 (11): 1067–1068.

65 Sirimanna, K.S., Todd, G.B., and Madden, G.J. (1992). A randomized study to compare calcium sodium alginate fiber with two commonly used materials for packing after nasal surgery. *Clin. Otolaryngol.* 17 (3): 237–239.

66 Sirimanna, K.S., Todd, G.B., and Madden, G.J. (1994). Early complications of packing after nasal surgery with three different materials. *Ceylon Med. J.* 39 (3): 129–131.

67 Skjak-Braek, G. and Espevik, T. (1996). Application of alginate gels in biotechnology and biomedicine. *Carbohydr. Eur.* 14: 19–25.

68 Smith, J. (1992). Comparing Sorbsan and polynoxylin/Melolin dressings after toenail removal. *J. Wound Care* 1 (3): 17–19.

69 Smith, J. and Lewis, J.D. (1990). Sorbsan and leg ulcers (letter). *Pharm. J.* 244: 468.

70 Stacey, M.C., Jopp-Mckay, A.G., Rashid, P. et al. (1997). The influence of dressings on venous ulcer healing-a randomized trial. *Eur. J. Vasc. Endovasc. Surg.* 13 (2): 174–179.

71 Steenfos, H.H. and Agren, M.S. (1998). A fiber-free alginate dressing in the treatment of split-thickness skin graft donor sites. *J. Eur. Acad. Dermatol. Venereol.* 11 (3): 252–256.

72 Suzuki, Y., Nishimura, Y., Tanihara, M. et al. (1998). Evaluation of a novel alginate gel dressing: cytotoxicity to fibroblasts in vitro and foreign-body reaction in pig skin in vivo. *J. Biomed. Mater. Res.* 39 (2): 317–322.

73 Thomas, S. (1985). Use of a calcium alginate dressing. *Pharm. J.* 235: 188–190.

74 Thomas, S. (1990). Sorbsan and leg ulcers (letter). *Pharm. J.* 244: 468–469.

75 Thomas, S. (1990). *Wound Management and Dressings*. London: The Pharmaceutical Press.

76 Thomas, S. (1992). Alginates. *J. Wound Care* 1 (1): 29–32.

77 Thomas, S. (2000). Alginate dressings in surgery and wound management, Part 1. *J. Wound Care* 9 (2): 56–60.

78 Thomas, S. (2000). Alginate dressings in surgery and wound management, Part 2. *J. Wound Care* 9 (3): 115–119.

79 Thomas, S. (2000). Alginate dressings in surgery and wound management, Part 3. *J. Wound Care* 9 (4): 163–166.

80 Thomas, S. and Tucker, C.A. (1989). Sorbsan in the management of leg ulcers. *Pharm. J.* 243: 706–709.

81 Thomas, S., Fisher, B., Fram, P.J. et al. (1998). Odour-absorbing dressings. *J. Wound Care* 7 (5): 246–250.

82 Turner, T.D. (1989). The development of wound management products. *Wounds* 1 (3): 155–171.

83 Van Gils, C.C., Roeder, B., Chesler, S.M. et al. (1998). Improved healing with a collagen alginate dressing in the chemical matricectomy. *J. Am. Podiatr. Med. Assoc.* 88 (9): 452–456.

84 Vanstraelen, P. (1992). Comparison of calcium sodium alginate (Kaltostat) and porcine xenograft (E-Z Derm) in the healing of split-thickness skin graft donor sites. *Burns* 18 (2): 145–148.

85 Varma, S.K., Henderson, H.P., and Hankins, C.L. (1991). Calcium alginate as a dressing for mini skin grafts (skin soup). *Br. J. Plast. Surg.* 44 (1): 55–56.

86 Wei, S., Jiang, C., Hou, M. et al. (2014). Effects of alginate dressing on perianal disease patients after operation. *Chin. J. Mod. Nurs.* 20 (5): 601–603.

87 Xie, M., Xu, G., Li, Y. et al. (2003). Comparison of the efficacy of four nasal packing materials. *China J. Endosc.* 9 (12): 19–22.

88 Xie, M., Xu, G., Li, Y. et al. (2003). Comparison of the clinical efficacy of four filling materials for nose. *China J. Endosc.* 9 (12): 19–22.

89 Young, M.J. (1993). The use of alginates in the management of exudating, infected wounds: case studies. *Dermatol. Nurs.* 5 (5): 359–363.

90 Young, M.J. (1993). The use of alginates in the management of exudating, infected wounds: case studies. *Dermatol. Nurs.* 5 (5): 356–363.

91 Zimmermann, U., Klock, G., Federlin, K. et al. (1992). Production of mitogen-contamination free alginates with variable ratios of mannuronic acid to guluronic acid by free flow electrophoresis. *Electrophoresis* 13: 269–274.

9

Functional Modifications of Alginate Fibers and Wound Dressings

9.1 Introduction

The production of wound dressings from alginate represents a chain of processes comprising the extraction of sodium alginate from brown seaweeds to wet spinning of alginate fibers, production of nonwoven fabrics, packaging and sterilization, etc. In each of these many steps, innovative technologies have been applied to modify the structure and properties so that the end product can better meet clinical requirements. These improvements aim to increase absorption capacities, enhance antimicrobial properties, and introduce many other novel features.

In addition to their own unique structures and product performances, alginate fibers and wound dressings can be combined with other types of materials to produce modified materials with enhanced functionalities. In this respect, the development in other related fields, such as nano-materials, biomaterials, micro-electronics, and biotechnology, has made it possible to develop smart materials for medical and healthcare applications by utilizing functional biomaterials, photochromic materials, thermochromic materials, conductive polymers, and fibers, as well as phase change materials, light-emitting polymers, optical fibers, shape memory polymers, etc.

As illustrated in Figure 9.1, functional modifications can be carried out in the production of sodium alginate raw materials, in the alginate fiber-making process, in the development of alginate fiber-based textile materials, and with novel applications of alginate-based wound dressings. The many developments in these areas are summarized in this chapter.

9.2 Chemical Modification of Alginic Acid

Alginate is a linear anionic polysaccharide consisting of two types of 1,4-linked hexuronic acid residues, namely, β-D-mannuronopyranosyl (M) and α-L-guluronopyranosyl (G) residues, arranged in blocks of repeating M residues (MM blocks), blocks of repeating G residues (GG blocks), and blocks of mixed M and G residues

Alginate Fibers and Wound Dressings: Seaweed Derived Natural Therapy, First Edition. Yimin Qin.

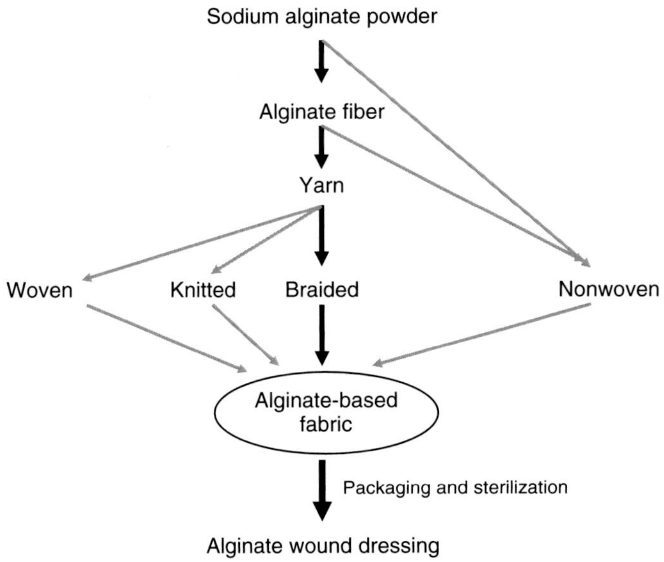

Figure 9.1 Structural hierarchy of alginate fiber-based wound dressings.

(MG blocks). Alginate has an abundance of free hydroxyl and carboxyl groups distributed along the polymer chain backbone, which can be chemically modified to generate derivatives with new functional characteristics. The chemical reactions include oxidation, reductive amination, sulfation, copolymerization, coupling of cyclodextrin units, and many others, which can be applied to the hydroxyl and carboxylic acid groups [107]. Through functionalizing available hydroxyl and carboxyl groups, the properties such as solubility, hydrophobicity, physicochemical, and biological characteristics can be modified to broaden applications [66]. When used in the fiber-making process, these chemically modified alginate derivatives can generate new properties for the alginate fibers and wound dressings.

9.2.1 Chemical Modification of the Hydroxyl Groups

9.2.1.1 Oxidation

Oxidation of alginate can generate more reactive groups and a faster degradation when used for drug delivery [11, 43]. Oxidation reaction is applied on the –OH groups at C-2 and C-3 positions of the uronic units of sodium alginate with sodium periodate, resulting in the rupture of carbon–carbon bond and the formation of two aldehyde groups in each oxidized monomeric unit, and as a result, larger rotational freedom and new reactive groups along the backbone are obtained. The degree of oxidation can be controlled by varying the concentration of the oxidant. Gomez et al. [32] synthesized oxidized alginates by using sodium periodate and found that when the degree of oxidation was over 10 mol%, no more gels were formed in the presence of excessive calcium ions. Figure 9.2 shows an illustration of the oxidation of sodium alginate by using sodium periodate.

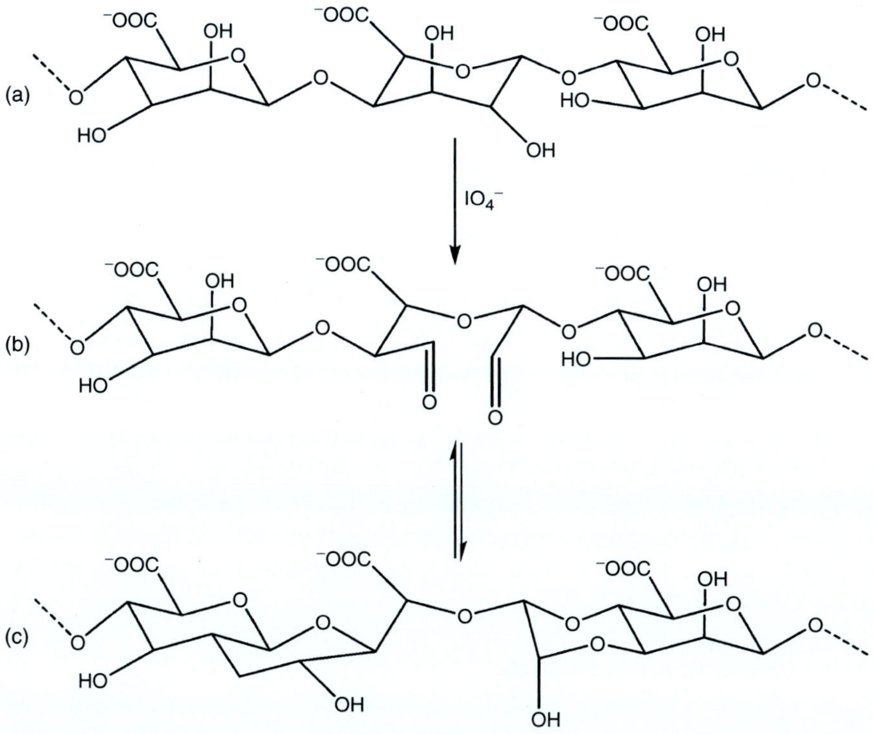

Figure 9.2 An illustration of the oxidation of sodium alginate by using sodium periodate. Source: Adapted with permission from He et al. [34].

9.2.1.2 Reductive-Amination of Oxidized Alginate

The aldehyde groups in oxidized alginate can be used to prepare more alginate derivatives by reductive amination, which can be carried out by using $NaCNBH_3$ as reducing agent, as it is more reactive and selective than the frequently employed sodium hydroborate ($NaBH_4$). The advantage of $NaCNBH_3$ is that the reduction of imine intermediate groups by $CNBH_3^-$ anion is rapid at pH values of 6–7, and the reduction of aldehyde or ketone is negligible in this pH range [4, 13, 39]. Kang et al. [39] used this method to prepare alginate-derived polymeric surfactants by the addition of long alkyl chains to the alginates, which endowed them with amphiphilic characteristics, such as lower surface tension, solubilizing of solid azobenzene, and adsorption of heavy metal ions in practical application. Li et al. [48] made microsphere beads with the alginate-derived polymeric surfactants in aqueous solution of sodium chloride and calcium chloride. A hydrophobic drug of ibuprofen was loaded on the modified alginate for controlling release *in vitro*. It was found that the loading level of drug was obviously increased and the release rate was well controlled. Laurienzo et al. [46] synthesized a novel alginate–polyethylene glycol (PEG) graft copolymer by reacting a mono-carboxyl terminated PEG with a sodium alginate modified by inserting a given amount of amine functionalities. The coupling between PEG and alginate was carried out using carbodiimide chemistry in aqueous solutions. The alginate–PEG copolymers retain the gelation characteristics of alginate since the PEG chemical grafting does not consume the carboxyl groups. The presence of grafted PEG molecules inside alginate gels can increase the pore dimensions and induce improved cell anchorage.

9.2.1.3 Sulfation

Sulfated alginate has high blood compatibility because of its structural similarity to heparin, which has been widely used for anticoagulant therapy [3]. Huang et al. [37] prepared alginate sulfate through reaction with $ClSO_3H$ in formamide using the following condition: 10 g sodium alginate is added to the sulfating reagent containing 80 ml formamide and 20 ml $ClSO_3H$, and the mixture is preserved at 60 °C for four hours to give a brown solution. In total, 200 ml acetone is added to precipitate the solution, and the precipitate is redissolved in distilled water and its pH is adjusted to 10–11 by 0.1 mol/l NaOH, then the solution is dialyzed for 72 hours and concentrated to give alginate sulfates. The *in vitro* coagulation assay of human plasma containing the sulfates indicated that alginate sulfates had considerably high anticoagulant activity especially in the intrinsic coagulation pathway. Figure 9.3 shows the preparation and chemical structure of sulfated alginate in the presence of dicyclohexylcarbodiimide.

9.2.1.4 Cyclodextrin-Linked Alginate

Cyclodextrin can be covalently linked to alginate to offer drug-carrying properties for controlled release. Pluemsab et al. [68] showed that the inclusion ability of alginate was introduced by covalently linking cyclodextrin to the hydroxyl groups of alginate. Stable and spherical beads were also obtained simply by dropping an aqueous solution of modified CD–alginate into a calcium chloride solution. Figure 9.4 shows the chemical structure of cyclodextrin.

9.2.1.5 Acetylation of Alginate

Schweiger [74] reported the synthesis of both partially and fully acetylated alginic acid derivatives using an acid-catalyzed esterification technique. The reaction was performed by suspending alginic acid in a mixture of acetic acid and acetic anhydride containing perchloric acid. Products with DS values up to 1.85 were achieved at moderate temperatures without significant degradation. Chamberlain et al. [15] found that the presence of water was essential to the availability of hydroxyl groups for

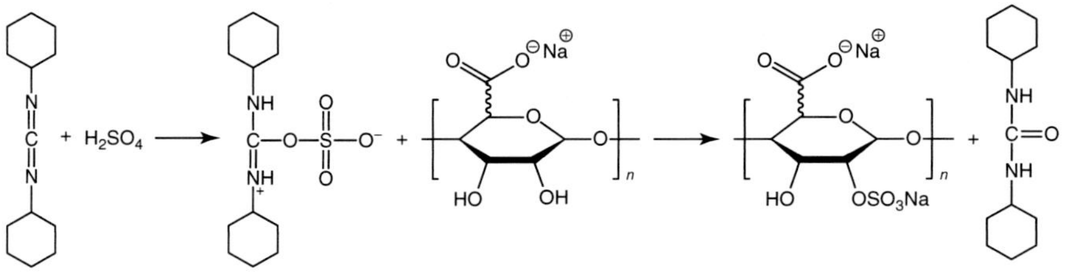

Figure 9.3 Preparation and chemical structure of sulfated alginate in the presence of dicyclohexylcarbodiimide.

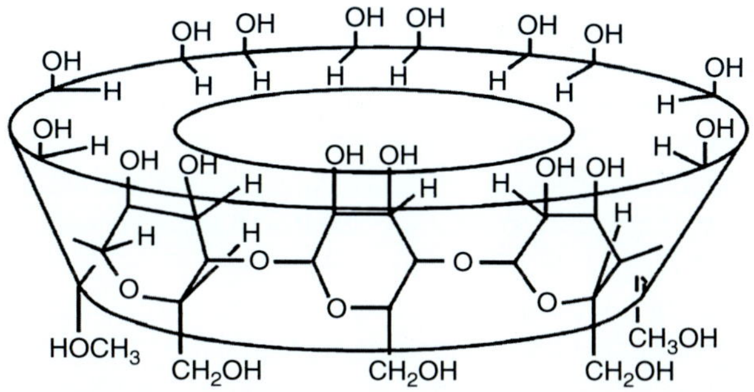

Figure 9.4 Chemical structure of cyclodextrin.

acetylation. Strong H bonds present in the dehydrated state prevented the hydroxyl groups from reacting. As the acetylation reaction progressed, H bonding between the vicinal hydroxyl groups was overcome and an alginate monoacetate product was formed, and there was a rapid increase in the reaction rate at DS values between 0.7 and 1.5. The increase in the reaction rate was associated with increasing availability of non-H-bonded hydroxyl groups for acetyl substitution.

Skjåk-Bræk et al. [81] reported the effects of acetylation upon alginate properties. Molecular weight measurements showed no significant degradation taking place during acetylation. The addition of acetyl groups to the backbone caused noticeable chain extension at 0.1 M ionic strength. Even a small proportion of acetyl groups on the backbone played an important role in defining the polymer conformation. When Ca-alginate gels were formed, acetylation was found to severely diminish the ability of Ca^{2+} ions to induce conformational ordering. Consequently, lower-strength gels resulted from acetylated alginates. On drying and re-swelling the Ca-alginate beads, the degree of swelling increased 500-fold from DS = 0 to DS = 0.65. This was a result of enhancement in the positive osmotic pressure resulting from a higher number of dissociating counter ions per polymer chain, due to impairment of co-operative binding by the added acetyl groups.

9.2.1.6 Phosphorylation of Alginates

Coleman et al. [20] prepared phosphorylated alginate derivatives by using urea/phosphoric acid reagent in a suspension of alginate in DMF. A maximum DS of 0.26 was achieved using an alginate:H_3PO_4:urea mole ratio of 1 : 20 : 70, whereby the heterogeneous nature of the reaction apparently limited the maximum DS attainable, and phosphoric acid as a strong acid is likely to cause alginate molecular weight degradation.

9.2.2 Chemical Modification of the Carboxyl Groups

9.2.2.1 Esterification

Alginate can be modified by direct esterification with several alcohols in the presence of catalyst, and the alcohol is present in excess to ensure that the equilibrium is in favor of product formation. Commercially, the only esterified derivative of alginate that is widely used is the propylene glycol alginate (PGA), which is obtained by esterification of alginate with propylene oxide. Yang et al. [105] carried out a reaction between the carboxylic acid groups of protonated sodium alginate and the hydroxyl of cholesterol at room temperature for 24 hours, after adding dicyclohexylcarbodiimide (DCC) as a coupling agent and 4-(*N*,*N*-dimethylamino) pyridine (DMAP) as a catalyst. Their research showed that the amphiphilic cholesteryl ester of alginate can self-assemble into the more stable and compact nano-aggregates through the intra- and inter-molecular hydrophobic interactions between cholesteryl grafts in aqueous NaCl solution, compared with the parent sodium alginate. Broderick et al. [12] reported that butyl ester of alginate prepared from sodium alginate through esterification with butanol in the presence of concentrated sulfuric acid as catalyst at room temperature for 18 hours. The study showed that the novel material is capable of encapsulating both hydrophilic and hydrophobic molecules.

Ester of alginate can be prepared from an alkyl halide with the carboxylic groups of alginate, previously transformed into their tetrabutylammonium (TBA) salts in homogeneous medium [7, 10, 67]. A simple procedure is summarized as follows: sodium alginate is first transformed into its acidic form by treatment with ethanolic HCl. After filtration, the resulting acidic polysaccharide is washed with ethanol (70%) until any remnant of chloride ion is removed, then with acetone. After drying at room temperature and reduced pressure, this compound is dispersed in water and neutralized (pH 7.0) by tetrabutylammonium hydroxide (TBA-OH) under controlled-delivery conditions. The TBA salt of alginic acid is dissolved in dimethylsulfoxide (DMSO). Then an alkyl halide (dodecyl bromide) is introduced at adequate stoichiometry and left to react for 24 hours under stirring at room temperature. The long alkyl chains are thus linked to the backbone of alginate chain via ester functions.

9.2.2.2 Amidation

Galant et al. [28] reported that alginate was hydrophobically modified by use of the coupling agent 1-ethyl-3-(3-dimethylaminopropyl) carbodiimide hydrochloride (EDC-HCl) to form amide linkages between amine-containing molecules and the carboxylate moieties on the alginate polymer backbone. The preparation of alginate derivatives via amide linkage has also been widely reported in the literature [31, 69, 104, 115]. This type of reaction is described as follows: an aqueous solution of sodium alginate is adjusted to pH 3.4 by the addition of HCl, then to this solution is added a certain amount of EDC-HCl. After five minutes of reaction, octylamine is added and the mixture is stirred for 24 hours at ambient temperature. The product is isolated by precipitation in acetone and the polymer is collected by filtration.

Abu-Rabeah et al. [1] reported that coupling of *N*-(3-aminopropyl) pyrrole to the alginate via amide linkages was achieved using the same method. They found that the pyrrole–alginate conjugate prepared at a degree of 30% of molar modification can be efficiently electropolymerized, providing a biocompatible host matrix that retained enzyme molecules by both gellification and electrochemical cross-linking. Vallee et al. [91] reported that new amphiphilic derivatives of sodium alginate were prepared by covalent attachment of dodecylamine onto the polysaccharide via amide linkages at different substitution ratios, using 2-chloro-1-methylpyridinium iodide (CMPI) as coupling reagent. The derivatives are prepared as follows: Na^+ alginate is transformed into its acidic form, and then neutralized to pH 7 by $TBA + OH^-$. The TBA-alginate salt is dissolved in DMF and stirred overnight to allow its complete dissolution. Then CMPI (required amount) and an excess of dodecylamine are added at 0 °C. Triethylamine is added into the solution at a concentration similar to that of dodecylamine. The solution is kept at 0 °C for 45 minutes, and then left at room temperature for 20 hours. Aqueous NaCl is added to the solution in order to exchange TBA^+ with Na^+ ions. Finally, the polymers are purified by precipitation in 7 : 1 EtOH–water, washed three times, and dried under diminished pressure.

9.2.3 Other Chemical Modifications

9.2.3.1 Organic Soluble Derivative of Alginate

The solubility of alginates in organic media requires formation of a TBA salt, with the complete dissolution of TBA–alginate in polar aprotic solvents containing tetrabutylammonium fluoride (TBAF) [65]. The solubility of alginates depends strongly on the state of the backbone carboxylic acid groups. Alginic acid with its carboxylic acid groups in their protonated form was not fully soluble in any solvent system examined, including water. Na-alginate dissolved in water but was not entirely soluble in any organic medium examined. TBA-alginate was completely soluble in water, ethylene glycol, and polar aprotic solvents containing TBAF but did not dissolve in any other solvent systems under consideration.

9.2.3.2 Attachment of Cell Signaling Molecules

Alginates are important biomaterials in the area of bioengineering [6]. A critical requirement of such biomaterials is their ability to provide an environment that is both physically and chemically favorable to the presence of biological species such as living cells. In order to enhance the chemical interactions of alginate matrices with cells, they are functionalized with cell-specific ligands or extracellular signaling molecules. In addition to enhancement of the cellular interactions, functionalization may also play a role in controlling the growth, differentiation, and behavior of cells in culture [80]. The ability of alginates to form gels capable of encapsulating cells, drugs, and other biological entities is another important advantage for biological applications. However, a possible drawback is that cells do not adhere to alginates naturally. Chemical functionalization with cell signaling moieties is therefore crucial in order to overcome the low affinity of alginates to cell surfaces. Donati et al. [22] reported the covalent attachment of galactose moieties to the alginate backbone for enhancement of hepatocyte cell recognition. Hepatocytes perform a variety of metabolic functions in the normal liver. However, outside the liver, they lose their function and are viable for only short time periods. The presence of an anchor material that provides mechanical support and immunoprotection to the cells prevents such loss of functionality and cell death.

Alginates with their unique gel-forming ability are considered very good matrix materials, providing a suitable environment for sustaining hepatocytes outside the organism. The surface of hepatocyte cell membranes possesses an asialoglycoprotein receptor (ASGPR), which binds and internalizes glycoproteins with terminal b-galactosyl residues. Galactose-modified alginates capable of ionic gelation were therefore synthesized to be used as biomaterials for improved encapsulation and adhesion of hepatocytes. The ability of galactose-substituted alginate derivatives to form gel beads was studied. Substitution of galactose sugar residues on the alginate backbone caused a reduction in the number of available carboxylic acid groups and impacted the bead size. Having galactose side chains caused conformational disordering within gels, whereby the junctions formed were held less tightly compared to unmodified alginates. A result of this effect was that larger bead volumes were achieved due to a net gain in hydration. Galactose moieties also interfered with the polymer chain packing process during gel formation. When the gel beads were dried and rehydrated, the kinetics and thermodynamics of swelling were affected. Larger volume increases during rehydration resulted in high DS values. An increase in the DS was also found to decrease the affinity of alginates toward Ca^{2+} ions.

9.2.3.3 Covalent Cross-linking of Alginates

Chemical cross-linking can be used to create stable and robust networks for the alginate molecules, which is useful in the preparation of hydrogels, foams, and fibers. Grasselli et al. [33] used chemical cross-linking of alginates to make beads for affinity and ion exchange chromatography. The cross-linking was achieved by reacting Ca-alginate beads with epichlorohydrin in the presence of aqueous NaOH. The cross-linked alginate beads were stable in high ionic strength media and polar solvents. In addition, large improvements in mechanical and chemical resistance properties were observed for covalently cross-linked alginates compared to Ca-alginate gels. Using similar procedure, Moe et al. [56, 57] prepared super-swelling alginate materials that were able to swell up to 100 times their dry volume without mass loss.

In addition to epichlorohydrin, alginate can also be cross-linked with glutaraldehyde. Yeom et al. [109] prepared sodium alginate films and carried out the cross-linking reaction in an acetone solution containing glutaraldehyde and HCl. These cross-linked alginates can be used to carry various functional materials [2, 8, 63, 64, 87].

When applied to alginate fibers, Kim et al. [40] showed that the use of glutaraldehyde cross-linked alginates can form superabsorbent fibers for use in disposable diapers and sanitary napkins. In addition, it can also be used to prepare thermodynamically controlled alginate network gels having pH-responsive properties [16].

Leone et al. [47] reported covalent cross-linking of alginate by amide bond formation. Na-alginate was first converted to TBA-alginate and then dissolved in DMF for reaction. 2-Chloro-*N*-methyl pyridinium iodide (CMPI) was used to activate the carboxyl groups. A reactive diamine was then used along with triethylamine catalyst to yield cross-linked alginates.

9.2.3.4 Graft Copolymerization of Alginates

Graft copolymerization can be applied on alginate to prepare a number of functional materials. For example, Liu et al. [50] prepared a superabsorbent from copolymer of sodium acrylate with sodium alginate, which can absorb 1000 and 85 times its own mass, respectively, in distilled water and 0.9% aqueous NaCl at room temperature. Graft reaction can take place at hydroxyl group. Sen et al. [75] reported the synthesis of various grades of graft copolymers based on acrylamide and sodium alginate via microwave irradiation. Their experiment showed that the copolymer having higher percentage of grafting and molecular weight is a better flocculant in coal suspension compared with other grades of the grafted polymer and sodium alginate. Sand et al. [72] reported an alginate-g-vinyl sulfonic acid prepared by employing potassium peroxydiphosphate/thiourea redox system. The synthesized graft copolymer shows better results for swelling, metal ion uptake, flocculating, and resistance to biodegradability properties in comparison to alginate.

Graft copolymerization is a powerful method for modification of the physical and chemical properties of alginates. Grafting synthetic polymer chains to the alginate backbone can introduce hydrophobicity and steric bulkiness, which help to protect the polysaccharide backbone from rapid dissolution and erosion. This in turn translates into a sustained release of active molecules from alginate matrices. Shah et al. [78] reported the ceric ammonium nitrate (CAN)-induced grafting of poly-acrylonitrile (PAN), poly-methyl acrylate (PMA), or poly-methyl methacrylate (PMMA) onto alginates. Homopolymer by-products were separated from the grafted alginates by solvent extraction. A comparison of the three different acrylate monomers showed that the grafting efficiency was highest for AN followed by MA followed by MMA. The homopolymer fractions formed showed a reverse trend where MMA polymerized to a greater extent compared to MA and AN. Alginate can also be grafted by using Fenton's reagent, another redox initiator system for grafting synthetic polymers onto alginates [76, 77].

Tripathy et al. [89, 90] described the grafting of poly-acrylamide (PAAm) onto alginates using a ceric-induced system. The advantage of using acrylamide (AAm) compared to AN, MA, or MMA for radical grafting using CAN was that the homopolymer PAAm did not form in the mixture. In the case of AAm, radical formation takes place only on the alginate backbone leading exclusively to grafted products. PAAm-grafted alginate (PAAm-g-Alg) was evaluated for its use as a commercial flocculant, where alginates containing longer PAAm grafts performed better than short grafts. Hydrolysis of the amide groups of PAAm-g-Alg was described through aqueous KOH treatment [9]. When in contact with KOH under certain controlled conditions, the amide groups of PAAm-g-Alg can hydrolyze without causing alginate degradation. Hydrolyzed PAAm-g-Alg possessed enhanced flocculation and thickening characteristics vs commercial flocculants.

Mollah et al. [58] described an approach where 3-(trimethoxysilyl) propyl methacrylate was grafted onto Na-alginate to form films through casting. The films were soaked in a silane monomer mixture and were subsequently photo-cross-linked to study the physico-mechanical properties of the networks formed. Kulkarni [45] described the synthesis of electrically responsive alginate-based hydrogels by grafting PAAm to the backbone and subsequently hydrolyzing the amide. Similar electro-responsive hydrogels were also described by Yang et al. [106] by grafting poly-acrylic acid (PAA) onto the alginate backbone. The derivatives were used as transdermal drug delivery systems for ketoprofen, a model nonsteroidal anti-inflammatory drug. On applying an electrical stimulus across the hydrogel, the positive counterion moved toward the negative electrode, but the negative carboxylate groups remained immobile. Consequently, an increased osmotic pressure near the positive electrode led to de-swelling of the hydrogel, causing release of ketoprofen from the matrix.

Omagari et al. [60] described a chemoenzymatic route to prepare amylose-grafted alginates, and disintegrable beads thereof. Amylose can be prepared by a phosphorylase-catalyzed enzymatic polymerization initiated from a maltooligosaccharide primer. Thus, an amine functional maltooligosaccharide primer was first synthesized and attached to the alginate backbone via aqueous EDC/NHS coupling. Phosphorylase-catalyzed polymerization was then performed to propagate amylose chains from the alginate backbone.

Kim et al. [41] described the grafting of poly-*N*-isopropylacrylamide (PNIPAAm) to the alginate backbone to make temperature/pH-responsive hydrogels. PNIPAAm hydrogels are well known for their ability to show lower critical solution temperature (LCST) behavior in aqueous solutions, while alginates show pH-responsive behavior due to the presence of carboxylic acid groups on the backbone. The temperature and pH-responsive properties of PNIPAAm and alginates were combined by a grafting technique. Amine terminal PNIPAAm (PNIPAAm-NH_2) was first synthesized and reacted with the carboxyl groups on the alginate backbone using EDC/NHS coupling.

Gao et al. [30] reported the grafting of poly-2-dimethylamino-ethyl methacrylate (PDMAEMA) onto oxidized Na-alginate (OAlg) to study the *in vitro* controlled release behavior of BSA protein from the hydrogels. PDMAEMA is a water-soluble polymer with tertiary amine groups containing side chains and exhibits LCST behavior. Grafting of PDMAEMA chains to the alginate chain allowed control over the equilibrium swelling ratio of hydrogels employing pH and ionic strength of the media. As the pH of the medium is increased above 3.0, ionization of the carboxylic acid groups takes place leading to the formation of polyelectrolyte complexes with the PDMAEMA cations. This causes a decrease in the equilibrium swelling ratio. Increasing the ionic strength (NaCl concentration) initially causes an increase in the swelling ratio due to disruption of the polyelectrolyte complexes, and later decreases the swelling ratio as the gel network is destroyed by Na—Ca ion exchange.

9.3 Innovations in the Fiber-Making Process

Over the last few decades, alginate fibers with enhanced functional performances have been developed by using ion exchange, polymer blending, and other novel processing techniques. As illustrated in Figure 9.5, through ion exchange with their respective salts, alginate fibers containing zinc, copper, barium, silver, aluminum, beryllium, and chromium ions can be prepared. By blending various materials into

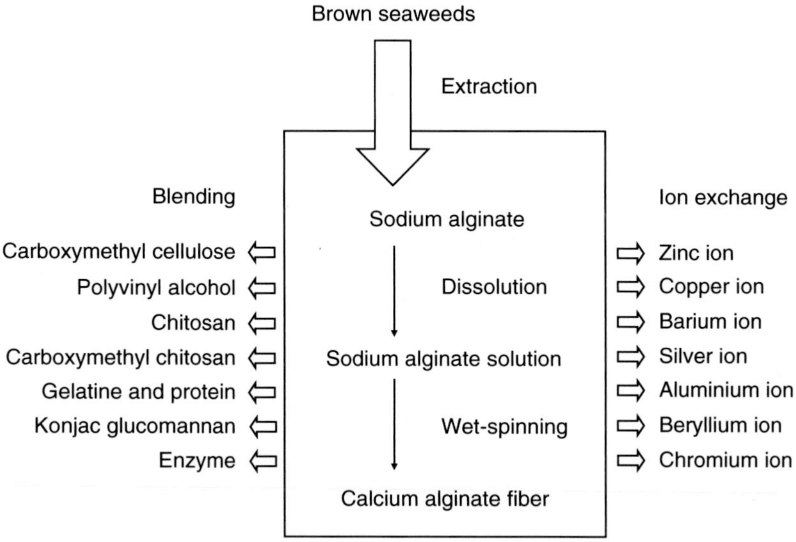

Figure 9.5 Schematic illustration of the various methods for the functional modifications of alginate fibers.

the sodium alginate spinning solution, alginate fibers can also be made to incorporate carboxymethyl cellulose, polyvinyl alcohol, chitosan, carboxymethyl chitosan, konjac, enzyme, and other bioactive materials. These fibers combine the properties of alginate and the various additives, making them useful in a large variety of applications.

9.3.1 The Production of Alginate Fibers Containing Metal Ions and Inorganic Compounds

In the early days for the development of alginate fibers, Speakman et al. [82] used ion-exchange process to prepare a series of alginate fibers containing different metal ions. The main purpose was to find a type of alginate that can be stable under alkaline conditions, unlike calcium alginate fibers which dissolve in such conditions. Chromium and beryllium alginate fibers were prepared by treating calcium alginate fibers or alginic acid fibers with chromium acetate and beryllium acetate, respectively. These fibers showed stability in alkaline conditions.

Many metal ions have special health-benefiting properties, with silver, zinc, and copper ions playing important roles in the wound healing process and in the development of wound management products. For example, copper alginate fibers possess excellent antimicrobial properties. As can be seen in Figure 9.6, when two strips of calcium alginate and copper alginate dressings were placed on Petri dishes with one side coated with bacteria, the copper alginate dressing was able to block the spreading of bacteria, while the bacteria were able to migrate along the calcium alginate strip to form colony.

Other functional properties of alginate fibers can also be improved through the use of selective metal ions. Barium alginate fibers can be prepared by extruding sodium

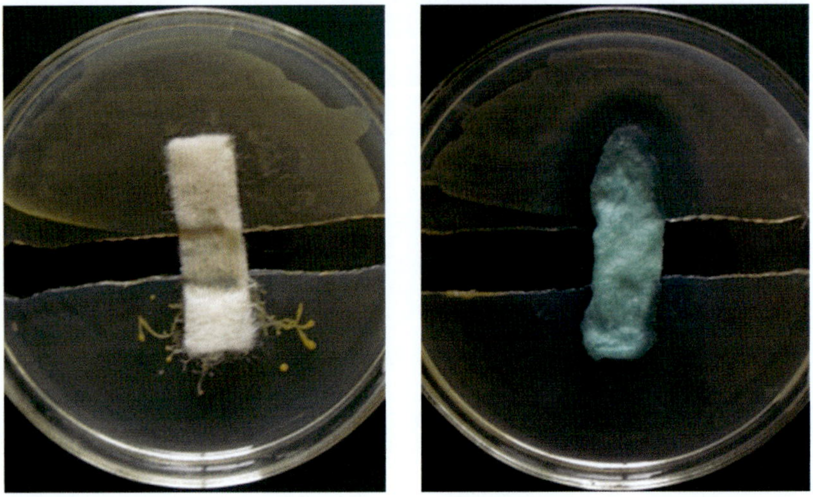

Figure 9.6 An illustration of the antimicrobial effect of calcium alginate and copper alginate nonwoven dressings.

alginate solution into a coagulation bath containing 0.5 mol/l barium chloride [96]. Test results showed that the average tensile strength of the barium alginate fiber was 20.69 cN/tex and the linear density was 3.86 dtex. Barium alginate fibers have excellent anti-radiation properties. Sodium alginate solution can also be extruded into aqueous aluminum chloride solution to form aluminum alginate fibers [5], which are stable in alkaline solutions.

Inorganic materials can also be loaded into alginate fibers by dispersing their powders in the sodium alginate spinning solution. When the diameter of these powders is below 1 μm, they can be easily dispersed in the spinning solution and extruded through the spinneret holes, which typically have a diameter of about 70 μm. Table 9.1 summarizes some of the alginate fibers containing inorganic materials.

9.3.2 The Production of Polyblend Fibers of Alginate and Other Polymers

Following the successful commercialization of alginate fibers containing sodium carboxymethyl cellulose, many attempts have been made to make polyblend fibers containing alginate and other polymers. Table 9.2 summarizes the various types of polyblend fibers and their properties.

9.3.3 The Production of Alginate Fibers Through Electrospinning

Electro-spinning is a straightforward method of generating ultrafine fibers on a nanoscopic scale with controlled surface morphology. Due to their high specific surface area and porous structure, the electrospun nonwoven fabrics that consist of

Table 9.1 Alginate fibers containing inorganic materials.

Inorganic material	Product performance
ZnO	Inhibit the growth of bacteria such as *E. coli* and *Staphylococcus aureus* [52]
AgCl	Increase the adsorption capacity for iodine [113]
$Al(OH)_3$	Improve thermal stability [114]
SiO_2	Improve flame retardancy [54]
Hydroxyapatite	Improve biocompatibility with osteoblasts [14]
Activated carbon	Enhance the adsorption of heavy metal ions and toxic organic matter [61]
Carbon nanotube	Increase the electrical conductivity and the adsorption for methylene blue, methyl orange, and other dyes [84, 92]
Graphene oxide	Improve mechanical properties of the fiber [35]

Table 9.2 Polyblend fibers of alginate and other polymers.

Polymer	Main properties of the polyblend fiber
Gelatine	Good biocompatibility and hygroscopicity [23–25, 94, 98, 110]
Polyvinyl alcohol	Good hygroscopicity and physicochemical properties [21, 83]
Konjac glucomannan	Improved moisture absorption [27]
Polyacrylonitrile	Sodium alginate was used to improve the skin affinity, dyeing property, and pilling performance of polyacrylonitrile fiber [29]
Cellulose	Sodium alginate was used to improve the hygroscopic property of the polyblend fiber [88, 111]
Pullulan	The hygroscopic property of the polyblend fiber was improved by the addition of pullulan [102]
Lecithin + poly-ethylene oxide	Good biocompatibility and water absorption [62]
Polyacrylamide	Improved dyeing properties of the fibers [103]
Krill protein	Improved biocompatibility of the polyblend fiber [108]

ultrafine fibers have found wide biomedical applications such as scaffolds for tissue engineering, wound dressings, and carriers for drug delivery. Chen et al. [19] successfully fabricated drug-loaded nanofibers using amphiphilic alginate derivative (AAD) and poly-vinyl alcohol (PVA) by electro-spinning. The AAD was synthesized by amide linkage attachment of octylamine onto the carboxylic group of alginate. The octyl groups of AAD could effectively make alginate chains flexible and enhance chain entanglements through the hydrophobic interactions between octyl groups.

In contrast to sodium alginate/PVA blend solutions, AAD/PVA blend solutions showed more excellent electro-spinnability. Additionally, the different volume ratios of sodium alginate or AAD to PVA played an important role in the formation and morphology of the electrospun blend nanofibers. Although the amidation of alginate could not fundamentally alter the spinnability of alginate, the content of alginate in the blend nanofibers was appropriately increased. The new generated drug-loaded AAD/PVA (20/80) blend nanofibers exhibited relatively continuous and uniform nanofibers with the average diameter of 191.72 nm. Compared to the drug-loaded sodium alginate/PVA (20/80) blend nanofibers, the drug-loaded AAD/PVA (20/80) blend nanofibers presented a sustained release performance with the slow release of about 76% *k*-cyhalothrin during the initial period of 10 hours.

Ku et al. [44] found that the ring opening of uronate unit by sodium periodate dramatically improved the electro-spinnability of alginate. The oxidation broke the linkage of C-2 and C-3 in urinate unit and formed ring opened structure, which weakened the inter/intra-molecular hydrogen bonding. The less rigid alginate chains decreased the solution viscosity but improved the chain-to-chain entanglement. As a result, the oxidized alginate showed better electro-spinnability, while the crystallinity and the thermal properties of the resultant fibers were decreased.

Xiao and Lim [102] prepared pullulan-alginate ultrafine fibers from aqueous polymer solutions using a free-surface electro-spinning method. Aqueous pullulan solution (10%, w/w) could be electrospun into fibers of 110 nm in diameter with a broad diameter distribution. By contrast, continuous and smooth fibers were formed when 0.8–1.6% (w/w) alginate was added to the 10% (w/w) pullulan solutions, producing thinner fibers ranging from 57 to 87 nm in diameter.

Hu and Yu [36] and Chang et al. [17] used dual-jet system to prepare alginate and chitosan microfibers, where the electrospinning process involved two spinnerets to blend the two polymers in the same fiber.

9.3.4 The Production of Alginate Fibers Containing Drugs

By loading various biologically active materials with pharmacological effect, functional alginate fibers can be produced with good biocompatibility, antibacterial, analgesic, hygroscopic, wound healing promotion, and many other novel bioactivities [112]. Mookhoek et al. [59] prepared alginate fibers containing discrete liquid-filled vacuoles after emulsifying the organic solution and dispersing it in sodium alginate aqueous solution. After the spinning process, the emulsion is dispersed in the fiber structure to obtain the vacuolar alginate fiber that can be used for the controlled delivery of healing agents. Wang et al. [95] added salicylic acid into the alginate and poly-ethylene glycol polyblend fiber and studied its drug-releasing performance. When the content of PEG was 5%, the fiber had the highest strength of 13.41 cN/tex and an elongation at break of 23.13%. With the increase of poly-ethylene glycol content, the fiber swelling property was also improved, with improved drug release properties. Polyblend fibers of alginate and starch have similar drug-release properties, and with the increase of starch content in the fiber, the drug-release properties are also improved [97].

Mikolajczyk et al. [55] added cefepime hydrochloride antibiotic into alginate fiber through its interaction with the carboxylic acid groups in the alginate fiber. Chen et al. [19] prepared polyblend fibers of alginate and polyvinyl alcohol by electrospinning and used the nanofibers to load drugs. In addition to drugs, alginate fibers can also be used to support living cells. Liu et al. [51] mixed vascular endothelial cells with sodium alginate solution and obtained alginate fibers embedded with these cells. Takei et al. [85] combined bovine carotid endothelial cells with alginate in a similar wet-spinning process. Experimental results showed that 95% of the cells maintained their biological activity during the spinning process.

9.3.5 The Production of Alginate and Chitosan Composite Fibers

Chitosan is a natural polymer with many bioactivities, such as its excellent antibacterial, hemostatic, and wound healing promotion properties. By blending alginate with chitosan in a composite fiber, it is possible to combine the unique characteristics of these two natural polymers to generate better functional performances [18, 49, 93]. With chitosan, the $-NH_2$ groups in its structure are protonated into positively charged $-NH_3^+$ under acidic dissolution conditions. These positively charged molecules are difficult to combine with the negatively charged sodium alginate in the same aqueous solution, hence it is difficult to obtain polyblend fibers in a similar way to the polyblend fibers made from alginate and carboxymethyl cellulose. However, since chitosan solution and alginate solution can precipitate each other through polyelectrolyte complex formation, many types of composite materials such as film and sponge can be prepared [73].

In the wet spinning process, chitosan and sodium alginate can be combined through positive and negative charge combinations in the coagulation bath [79]. Tamura et al. [86] obtained the composite fiber by extruding sodium alginate solution into a coagulation bath containing chitosan. Knill et al. [42] used a similar method by first reducing the molecular weight of chitosan and adding it into the coagulation bath. It was found that the low molecular weight chitosan can penetrate into calcium alginate fibers better, resulting in improved antibacterial properties of the fibers. Majima et al. [53] extruded sodium alginate solution into a first coagulation bath containing 0.1% chitosan and then a second bath containing 3% $CaCl_2$. The resultant alginate/chitosan polyblend fibers have good adhesion to fibroblasts and are suitable for tissue engineering substrates.

In order to prepare a uniform spinning solution, Watthanaphanit et al. [99, 101] dispersed nano-whiskers of chitosan into the aqueous sodium alginate solution and obtained a composite fiber containing 0.05–2.00% w/w of chitosan. In another study [100], chitosan solution was first dispersed in an organic solution to form an emulsion before sodium alginate solution was added to form the spinning solution. Calcium alginate fibers containing chitosan particles were then obtained by wet spinning.

Qin [71] invented a method of preparing a uniform spinning solution from chitosan and alginate by blending acid-resistant PGA with chitosan solution, which resulted in a homogeneous and stable spinning solution containing these

Table 9.3 Swelling ratio of PGA/chitosan polyblend fiber.

	Swelling ratio (g/g)	
Ratio between PGA and chitosan (g/g)	**In deionized water**	**In normal saline**
0 : 100	2.91	2.88
5 : 100	3.45	3.10
10 : 100	4.57	4.15
20 : 100	5.78	4.65
30 : 100	7.65	5.22

two polymers. PGA is an esterified derivative obtained by the reaction of alginate and propylene oxide. In the fiber-forming process, the ester group in PGA reacts with the amine group in chitosan to form a stable cross-linked structure through the amide group, while some of the ester groups hydrolyze to generate sodium carboxylic acid groups that can hold a large amount of water. Table 9.3 shows the swelling ratio of PGA and chitosan polyblend fibers in deionized water and normal saline, with the swelling ratio increasing significantly with the increase of PGA content.

Chitosan can be chemically modified to obtain water-soluble derivatives that can be mixed with sodium alginate in the spinning solution. By treating chitosan with chloroacetic acid, ethylene oxide, and propylene oxide, it is possible to obtain carboxymethyl chitosan, hydroxyethyl chitosan, and hydroxypropyl chitosan which can form uniform spinning solutions with sodium alginate. Fan et al. [25, 26] prepared polyblend fibers containing 10–70% carboxymethyl chitosan after blending sodium alginate solution with carboxymethyl chitosan solution. Results showed that these two components in the blend system have good compatibility. When the content of carboxymethyl chitosan was 30%, the dry tensile strength of the polyblend fiber reached a maximum of 13.8 cN/tex. Jiang et al. [38] showed that the addition of carboxymethyl chitosan to calcium alginate fiber can significantly improve the water absorption of the fiber.

With alginate, the C—C bond in its monomers can be oxidized by a strong oxidizing agent to generate two aldehyde groups. The resultant oxidized alginate has cross-linking properties similar to formaldehyde and glutaraldehyde [66]. Qin [70] treated chitosan fibers with an aqueous solution of oxidized sodium alginate and successfully coated a layer of alginate on the surface of chitosan fibers, mainly through the reaction between the amine groups in chitosan and the aldehyde groups in the oxidized sodium alginate, as shown in Figure 9.7. The composite fibers obtained by Schiff bond can combine the antibacterial activity of chitosan and the hydrophilicity of sodium alginate.

Table 9.4 shows the absorption properties of chitosan nonwoven fabric modified with oxidized sodium alginate. It is clear that the coating of chitosan fibers with

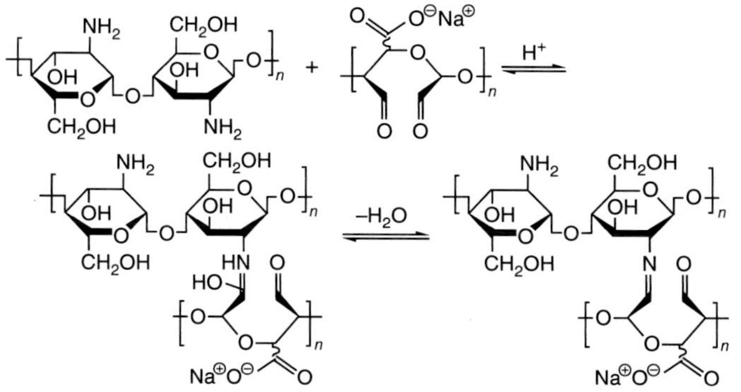

Figure 9.7 Schiff bond formation between oxidized sodium alginate and chitosan.

Table 9.4 Absorption properties of chitosan nonwoven fabric modified with oxidized sodium alginate.

Ratio between oxidized sodium alginate and chitosan nonwoven fabric (g/g)	Fiber swelling in deionized water (g/g)	Absorbency of the nonwoven fabric (g/g)
0 : 100	2.597	4.73
1 : 100	2.834	11.27
2.5 : 100	3.361	11.43
5 : 100	3.484	11.86
10 : 100	3.863	11.41
20 : 100	4.554	12.57
50 : 100	5.216	13.24

oxidized sodium alginate resulted in a significant improvement of the absorption properties of chitosan nonwoven fabric.

9.4 Summary

Alginate fibers and wound dressings can be modified in a number of ways. With alginate itself, a significant amount of research has been aimed at modifying alginate chemically, due to the free hydroxyl and carboxyl groups distributed along the backbone, to alter the characteristics in comparison to the native alginates. During the production of alginate fibers, the spinning solution is an aqueous solution of sodium alginate and by adding water-soluble polymers such as sodium carboxymethyl cellulose and inorganic compounds such as nano-silver antibacterial materials, alginate-based polyblend fibers can be prepared by wet spinning. Through the use of various bioactive substances, the properties of alginate fibers can be improved to make them better materials for the production of functional wound dressings.

References

1 Abu-Rabeah, K., Polyak, B., Ionescu, R.E. et al. (2005). Synthesis and characterization of a pyrrole–alginate conjugate and its application in a biosensor construction. *Biomacromolecules* 6: 3313–3318.

2 Adoor, S.G., Manjeshwar, L.S., Bhat, S.D. et al. (2008). Aluminum-rich zeolite beta incorporated sodium alginate mixed matrix membranes for pervaporation dehydration and esterification of ethanol and acetic acid. *J. Membr. Sci.* 318: 233–246.

3 Alban, S., Schauerte, A., and Franz, G. (2002). Anticoagulant activity polysaccharides: part I. Synthesis and structure–activity relationships of new pullulan sulfates. *Carbohydr. Polym.* 47: 267–276.

4 Andresen, I.L., Painter, T., and Smidsrod, O. (1977). Concerning the effect of periodate oxidation upon the intrinsic viscosity of alginate. *Carbohydr. Res.* 59: 563–566.

5 Atkinson W T. *The production of aluminium alginate fibers*. US Patent 3455697, 1969

6 Augst, A.D., Kong, H.J., and Mooney, D.J. (2006). Alginate hydrogels as biomaterials. *Macromol. Biosci.* 6: 623–633.

7 Babak, V.G., Skotnikova, E.A., Lukina, I.G. et al. (2000). Hydrophobically associating alginate derivatives: surface tension properties of their mixed aqueous solutions with oppositely charged surfactants. *J. Colloid Interface Sci.* 225: 505–510.

8 Bhat, S.D., Mallikarjuna, N.N., and Aminabhavi, T.M. (2006). Microporous alumino-phosphate molecular sieve-loaded novel sodium alginate composite membranes for pervaporation dehydration of aqueous-organic mixtures near their azeotropic compositions. *J. Membr. Sci.* 282: 473–483.

9 Biswal, D.R. and Singh, R.P. (2004). The flocculation and rheological characteristics of hydrolyzed and unhydrolyzed grafted sodium alginate in aqueous solutions. *J. Appl. Polym. Sci.* 94: 1480–1488.

10 Boisseson, M.R.D., Leonard, M., Hubert, P. et al. (2004). Physical alginate hydrogels based on hydrophobic or dual hydrophobic/ionic interactions: bead formation, structure, and stability. *J. Colloid Interface Sci.* 273: 131–139.

11 Boontheekul, T., Kong, H., and Mooney, D.J. (2005). Controlling alginate gels degradation utilizing partial oxidation and bimodal molecular weight distribution. *Biomaterials* 26: 2455–2465.

12 Broderick, E., Lyons, H., Pembroke, T. et al. (2006). The characterization of a novel covalently modified, amphiphilic alginate derivative, which retains gelling and non-toxic properties. *J. Colloid Interface Sci.* 298: 154–161.

13 Carre, M.C., Delestre, C., Hubert, P. et al. (1991). Covalent coupling of a short polyether on sodium alginate: synthesis and characterization of the resulting amphiphilic derivative. *Carbohydr. Polym.* 16: 367–379.

14 Chae, T., Yang, H., Leung, V. et al. (2013). Novel biomimetic hydroxyapatite/alginate nanocomposite fibrous scaffolds for bone tissue regeneration. *J. Mater. Sci. Mater. Med.* 24: 1885–1894.

15 Chamberlain, N.H., Cunningham, G.E., and Speakman, J.B. (1946). Alginic acid diacetate. *Nature* 158: 553.

16 Chan, A.W., Whitney, R.A., and Neufeld, R.J. (2009). Semi-synthesis of a controlled stimuli responsive alginate hydrogel. *Biomacromolecules* 10: 609–616.

17 Chang, J.J., Lee, Y.H., Wu, M.H. et al. (2012). Electrospun anti-adhesion barrier made of chitosan alginate for reducing peritoneal adhesions. *Carbohydr. Polym.* 88: 1304–1312.

18 Chang, J.J., Lee, Y.H., Wu, M.H. et al. (2012). Preparation of electrospun alginate fibers with chitosan sheath. *Carbohydr. Polym.* 87: 2357–2361.

19 Chen, X., Yan, H., Sun, W. et al. (2015). Synthesis of amphiphilic alginate derivatives and electrospinning blend nanofibers: a novel hydrophobic drug carrier. *Polym. Bull.* 72: 3097–3117.

20 Coleman, R.J., Lawrie, G., Lambert, L.K. et al. (2011). Phosphorylation of alginate: synthesis, characterization, and evaluation of in vitro mineralization capacity. *Biomacromolecules* 12: 889–897.

21 Cui, L., Liu, H., Rao, Y. et al. (2011). Electrospinning process and properties of sodium alginate polyvinyl alcohol solution. *China Synth. Fiber Ind.* 34 (3): 5–7.

22 Donati, I., Vetere, A., Gamini, A. et al. (2003). Galactose substituted alginate 1: preliminary characterization and study of gelling properties. *Biomacromolecules* 4: 624–631.

23 Du, Y., Fan, L., and Zhang, B. (2005). Preparation method and applications of alginate/gelatin fiber. Chinese Patent, ZL200510018615.8.

24 Fan, L., Du, Y., Huang, R. et al. (2005). Preparation and characterization of alginate/gelatin blend fibers. *J. Appl. Polym. Sci.* 96: 1625–1629.

25 Fan, L., Du, Y., Zhang, B. et al. (2005). Antibacterial fibers made of calcium alginate/chitosan derivative composite. *J. Funct. Polym.* 18 (3): 34–38.

26 Fan, L., Du, Y., Zhang, B. et al. (2006). Preparation and properties of alginate/carboxymethyl chitosan blend fibers. *Carbohydr. Polym.* 65: 447–452.

27 Fan, L., Zhu, H., Zheng, H. et al. (2007). Structure and properties of blend fibers prepared from alginate and konjac glucomannan. *J. Appl. Polym. Sci.* 106: 3903–3907.

28 Galant, C., Kjoniksen, A.L., Nguyen, G.T.M. et al. (2006). Altering associations in aqueous solutions of a hydrophobically modified alginate in the presence of cyclodextrin monomers. *J. Phys. Chem. B* 110: 190–195.

29 Gao, K., Li, H., and Shen, X. (2005). Preparation method for polyacrylonitrile and alginate blend fiber. Chinese Patent, CN1696361A.

30 Gao, C., Liu, M., Chen, S. et al. (2009). Preparation of oxidized sodium alginate graft-poly-2-dimethylamino ethyl methacrylate gel beads and in vitro controlled release behavior of BSA. *Int. J. Pharm.* 371: 16–24.

31 Gomez, C.G., Chambat, G., Heyraud, A. et al. (2006). Synthesis and characterization of a CD–alginate conjugate. *Polymer* 47: 8509–8516.

32 Gomez, C.G., Rinaudo, M., and Villar, M.A. (2007). Oxidation of sodium alginate and characterization of the oxidized derivatives. *Carbohydr. Polym.* 67: 296–304.

33 Grasselli, M., Diaz, L.E., and Cascone, O. (1993). Beaded matrices from cross-linked alginate for affinity and ion exchange chromatography of proteins. *Biotechnol. Tech.* 7: 707–712.

34 He, S.L., Zhang, M., Geng, Z. et al. (2005). Preparation and characterization of partially oxidized sodium alginate. *Chinese Journal of Applied Chemistry* 22: 1007–1011.

35 He, Y., Zhang, N., Gong, Q. et al. (2012). Alginate/graphene oxide fibers with enhanced mechanical strength prepared by wet spinning. *Carbohydr. Polym.* 88: 1100–1108.

36 Hu, W.W. and Yu, H.N. (2013). Coelectrospinning of chitosan/alginate fibers by dual-jet system for modulating material surfaces. *Carbohydr. Polym.* 95: 716–727.

37 Huang, R.H., Du, Y.M., and Yang, J.H. (2003). Preparation and in vitro anticoagulant activities of alginate sulfate and its quaterized derivatives. *Carbohydr. Polym.* 52: 19–24.

38 Jiang, L., Kong, Q., Ji, Q. et al. (2007). Study on preparation and properties of sodium alginate/water-soluble chitin blend fiber. *Synt Fiber China* (12): 12–15.

39 Kang, H., Jeon, G., Lee, M. et al. (2002). Effectiveness test of alginate-derived polymeric surfactants. *J. Chem. Technol. Biotechnol.* 77: 205–210.

40 Kim, Y.J., Yoon, K.J., and Ko, S.W. (2000). Preparation and properties of alginate superabsorbent filament fibers crosslinked with glutaraldehyde. *J. Appl. Polym. Sci.* 78: 1797–1804.

41 Kim, J.H., Lee, S.B., Kim, S.J. et al. (2002). Rapid temperature/pH response of porous alginate-g-poly(N-isopropylacrylamide) hydrogels. *Polymer* 43: 7549–7458.

42 Knill, C.J., Kennedy, J.F., Mistry, J. et al. (2004). Alginate fibres modified with unhydrolysed and hydrolysed chitosans for wound dressings. *Carbohydr. Polym.* 55 (1): 65–76.

43 Kong, H.J., Kaigler, D., and Mooney, D.J. (2004). Controlling rigidity and degradation of alginate hydrogels via molecular weight distribution. *Biomacromolecules* 5: 1720–1727.

44 Ku, M.K., Ahn, Y., Song, Y. et al. (2014). Effect of oxidized alginate on its electrospinnability. *Fibers Polym.* 15 (9): 1835–1841.

45 Kulkarni, R.V., Setty, C.M., and Sa, B. (2010). Polyacrylamide-g-alginate-based electrically responsive hydrogel for drug delivery application: synthesis, characterization, and formulation development. *J. Appl. Polym. Sci.* 115: 1180–1188.

46 Laurienzo, P., Malinconico, M., Motta, A. et al. (2005). Synthesis and characterization of a novel alginate–poly (ethylene glycol) graft copolymer. *Carbohydr. Polym.* 62: 274–282.

47 Leone, G., Torricelli, P., Chiumiento, A. et al. (2008). Amidic alginate hydrogel for nucleus pulposus replacement. *J. Biomed. Mater. Res. A* 84A: 391–401.

48 Li, Z., Ni, C., Xiong, C. et al. (2009). Preparation and drug release of hydrophobically modified alginate. *Chemistry* 1: 93–96.

49 Li, X., Xie, H., Lin, J. et al. (2009). Characterization and biodegradation of chitosan-alginate polyelectrolyte complexes. *Polym. Degrad. Stab.* 94: 1–6.

50 Liu, M.Z. and Cao, L.X. (2002). Preparation of a superabsorbent resistant to saline solution by copolymerization of acrylic acid with sodium polymannuronate. *Chin. J. Appl. Chem.* 19: 455–458.

51 Liu, Y., Sakai, S., and Taya, M. (2012). Production of endothelial cell-enclosing alginate-based hydrogel fibers with a cell adhesive surface through simultaneous cross-linking by horseradish peroxidase-catalyzed reaction in a hydrodynamic spinning process. *J. Biosci. Bioeng.* 114 (3): 353–359.

52 Lu, K. (2010). The preparation and antibacterial property of alginate fibers supported zinc oxide nanoparticles. *Mater. Res. Appl.* 4 (2): 125–127.

53 Majima, T., Funakosi, T., Iwasaki, N. et al. (2005). Alginate and chitosan polyion complex hybrid fibers for scaffolds in ligament and tendon tissue engineering. *J. Orthop. Sci.* 10: 302–307.

54 Mikolajczyk, T., Urbaniak-Domagala, W., and Wolowska-Czapnik, D. (2006). Effects of the structure of polymer and nanosilica additive on the sorption and electric properties of various alginate fibers. *J. Appl. Polym. Sci.* 101: 686–694.

55 Mikolajczyk, T., Wolowska-Czapnik, D., and Bogun, M. (2008). A new generation of fibers from alginic acid for dressing materials. *J. Appl. Polym. Sci.* 107: 1670–1677.

56 Moe, S.T., Skjåk-Bræk, G., and Smidsrød, O. (1991). Covalently cross-linked sodium alginate beads. *Food Hydrocoll.* 5: 119–123.

57 Moe, S.T., Skjåk-Bræk, G., Elgsaeter, A. et al. (1993). Swelling of covalently crosslinked alginate gels: influence of ionic solutes and nonpolar solvents. *Macromolecules* 26: 3589–3597.

58 Mollah, M.Z.I., Khan, M.A., Hoque, M.A. et al. (2008). Studies of physico-mechanical properties of photo-cured sodium alginate with silane monomer. *Carbohydr. Polym.* 72: 349–355.

59 Mookhoek, S.D., Fischer, H.R., and Van Der Zwaag, S. (2012). Alginate fibres containing discrete liquid filled vacuoles for controlled delivery of healing agents in fibre reinforced composites. *Compos. Part A* 43: 2176–2182.

60 Omagari, Y., Kaneko, Y., and Kadokawa, J.I. (2010). Chemoenzymatic synthesis of amylose grafted alginate and its formation of enzymatic disintegratable beads. *Carbohydr. Polym.* 82: 394–400.

61 Park, H.G., Kim, T.W., Chae, M.Y. et al. (2007). Activated carbon-containing alginate adsorbent for the simultaneous removal of heavy metals and toxic organics. *Process Biochem.* 42: 1371–1377.

62 Park, S.A., Park, K.E., and Kim, W.D. (2010). Preparation of sodium alginate/poly-ethylene oxide blend nanofibers with lecithin. *Macromol. Res.* 18 (9): 891–896.

63 Patil, M.B., Veerapur, R.S., Patil, S.A. et al. (2007). Preparation and characterization of filled matrix membranes of sodium alginate incorporated with aluminum-containing mesoporous silica for pervaporation dehydration of alcohols. *Sep. Purif. Technol.* 54: 34–43.

64 Patil, M.B., Veerapur, R.S., Bhat, S.D. et al. (2009). Hybrid composite membranes of sodium alginate for pervaporation dehydration of 1,4-dioxane and tetrahydrofuran. *Desalin. Water Treat.* 3: 11–20.

65 Pawar, S.N. and Edgar, K.J. (2011). Chemical modification of alginates in organic solvent systems. *Biomacromolecules* 12: 4095–4103.

66 Pawar, S.N. and Edgar, K.J. (2012). Alginate derivatization: a review of chemistry, properties and applications. *Biomaterials* 33: 3279–3305.

67 Pelletier, S., Hubert, P., Lapicque, F. et al. (2000). Amphiphilic derivatives of sodium alginate and hyaluronate: synthesis and physico-chemical properties of aqueous dilute solutions. *Carbohydr. Polym.* 43: 343–349.

68 Pluemsab, W., Sakairi, N., and Furuike, T. (2005). Synthesis and inclusion property of cyclodextrin-linked alginate. *Polymer* 46: 9778–9783.

69 Polyak, B., Geresh, S., and Marks, R.S. (2004). Synthesis and characterization of a biotin–alginate conjugate and its application in a biosensor construction. *Biomacromolecules* 5: 389–396.

70 Qin, Y. (2014). Preparation method and applications of chitosan fibers modified by oxidized alginate. Chinese Patent Zl201310127978.X.

71 Qin, Y. (2015). Preparation Methods and Applications of Polyblend Fibers From Chitosan and Propylene Glycol Alginate. Chinese Patent, Zl201210553968.8.

72 Sand, A., Yadav, M., and Behari, K. (2010). Synthesis and characterization of alginate-vinyl sulfonic acid with a potassium peroxydiphosphate/thiourea system. *J. Appl. Polym. Sci.* 118: 3685–3694.

73 Sather, H.V., Holme, H.K., Maurstad, G. et al. (2008). Polyelectrolyte complex formation using alginate and chitosan. *Carbohydr. Polym.* 74: 813–821.

74 Schweiger, R.G. (1962). Acetylation of alginic acid. II. Reaction of algin acetates with calcium and other divalent ions. *J. Org. Chem.* 27: 1789–1791.

75 Sen, G., Singh, R.P., and Pal, S. (2010). Microwave-initiated synthesis of polyacrylamide grafted sodium alginate: synthesis and characterization. *J. Appl. Polym. Sci.* 115: 63–71.

76 Shah, S.B., Patel, C.P., and Trivedi, H.C. (1994). Fenton's reagent-initiated graft copolymerization of acrylonitrile onto sodium alginate. *J. Appl. Polym. Sci.* 51: 1421–1426.

77 Shah, S.B., Patel, C.P., and Trivedi, H.C. (1994). Kinetics and reaction mechanism of Fenton's reagent-initiated graft copolymerization of acrylonitrile onto sodium alginate. *J. Appl. Polym. Sci.* 52: 857–860.

78 Shah, S.B., Patel, C.P., and Trivedi, H.C. (1995). Ceric-induced grafting of acrylate monomers onto sodium alginate. *Carbohydr. Polym.* 26: 61–67.

79 Shao, X. and Hunter, C.J. (2007). Developing an alginate/chitosan hybrid fiber scaffold for annulus fibrosus cells. *J. Biomed. Mater. Res.* 82A: 701–710.

80 Shapiro, L. and Cohen, S. (1997). Novel alginate sponges for cell culture and transplantation. *Biomaterials* 18: 583–590.

81 Skjåk-Bræk, G., Zanetti, F., and Paoletti, S. (1989). Effect of acetylation on some solution and gelling properties of alginates. *Carbohydr. Res.* 185: 131–138.

82 Speakman, J.B. and Chamberlain, N.H. (1944). The production of rayon from alginic acid. *J. Soc. Dye. Colour.* 60: 264–272.

83 Stone, S.A., Gosavi, P., Athauda, T.J. et al. (2013). In situ citric acid crosslinking of alginate/polyvinylalcohol electrospun nano fibers. *Mater. Lett.* 112: 32–35.

84 Sui, K., Li, Y., Liu, R. et al. (2012). Biocomposite fiber of calcium alginate multi-walled carbon nanotubes with enhanced adsorption properties for ionic dyes. *Carbohydr. Polym.* 90: 399–406.

85 Takei, T., Sakai, S., Ijima, H. et al. (2006). Development of mammalian cell-enclosing calcium-alginate hydrogel fibers in a co-flowing stream. *Biotechnol. J.* 1: 1014–1017.

86 Tamura, H., Tsuruta, Y., and Tokura, S. (2002). Preparation of chitosan-coated alginate filament. *Mater. Sci. Eng. C* 20: 143–147.

87 Teli, S.B., Gokavi, G.S., Sairam, M. et al. (2007). Highly water selective silicotungstic acid incorporated novel sodium alginate hybrid composite membranes for pervaporation dehydration of acetic acid. *Sep. Purif. Technol.* 54: 178–186.

88 Tian, S., Liu, H., and Wang, L. (2006). Preparation method for cellulose alginate blend fiber. Chinese Patent, CN1858312A.

89 Tripathy, T. and Singh, R.P. (2001). Characterization of poly-acrylamide grafted sodium alginate: a novel polymeric flocculant. *J. Appl. Polym. Sci.* 81: 3296–8308.

90 Tripathy, T., Pandey, S.R., Karmakar, N.C. et al. (1999). Novel flocculating agent based on sodium alginate and acrylamide. *Eur. Polym. J.* 35: 2057–2072.

91 Vallee, F., Muller, C., Durand, A. et al. (2009). Synthesis and rheological properties of hydrogels based on amphiphilic alginate–amide derivatives. *Carbohydr. Res.* 344: 223–228.

92 Vijoya, S., Konstantin, G., and Kornev, A. (2011). A method for wet spinning of alginate fibers with a high concentration of single-walled carbon nanotubes. *Carbon* 49: 1859–1868.

93 Wan, A., Liao, I., Yim, E. et al. (2004). Mechanism of fiber formation by interfacial polyelectrolyte complexation. *Macromolecules* 37: 7019–7025.

94 Wang, Q., Du, Y., Hu, X. et al. (2006). Preparation of alginate/soy protein isolate blend fibers through a novel coagulating bath. *J. Appl. Polym. Sci.* 101: 425–431.

95 Wang, Q., Zhang, N., Hu, X. et al. (2007). Alginate/polyethylene glycol blend fibers and their properties for drug controlled release. *J. Biomed. Mater. Res.* 82A: 122–128.

96 Wang, B., Kong, Q., Ji, Q. et al. (2009). The production of barium alginate fibers. *Funct. Mater.* 2 (40): 345–347.

97 Wang, Q., Hu, X., Du, Y. et al. (2010). Alginate/starch blend fibers and their properties for drug controlled release. *Carbohydr. Polym.* 82: 842–847.

98 Wang, Q.Q., Liu, Y., Zhang, C.J. et al. (2019). Alginate/gelatin blended hydrogel fibers cross-linked by Ca^{2+} and oxidized starch: preparation and properties. *Mater. Sci. Eng. C* 99: 1469–1476.

99 Watthanaphanit, A., Supaphol, P., Tamura, H. et al. (2008). Fabrication, structure, and properties of chitin whisker-reinforced alginate nanocomposite fibers. *J. Appl. Polym. Sci.* 110: 890–899.

100 Watthanaphanit, A., Supaphol, P., Furuike, T. et al. (2009). Novel chitosan-spotted alginate fibers from wet-spinning of alginate solutions containing emulsified chitosan-citrate complex and their characterization. *Biomacromolecules* 10 (2): 320–327.

101 Watthanaphanit, A., Supaphol, P., Tamura, H. et al. (2010). Wet-spun alginate/chitosan whiskers nanocomposite fibers: preparation, characterization and release characteristic of the whiskers. *Carbohydr. Polym.* 79: 738–746.

102 Xiao, Q. and Lim, L.T. (2018). Pullulan-alginate fibers produced using free surface electrospinning. *Int. J. Biol. Macromol.* 112: 809–817.

103 Xie, D. and Huang, T. (2019). Preparation and characterization of alginate and polyacrylamide blend fiber. *Artif. Fiber* 49 (4): 2–5.

104 Yang, J., Goto, M., Ise, H. et al. (2002). Galactosylated alginate as a scaffold for hepatocytes entrapment. *Biomaterials* 23: 471–479.

105 Yang, L.Q., Zhang, B.F., and Wen, L.Q. (2007). Amphiphilic cholesteryl grafted sodium alginate derivative: synthesis and self-assembly in aqueous solution. *Carbohydr. Polym.* 68: 218–225.

106 Yang, S., Liu, G., Cheng, Y. et al. (2009). Electro-responsive behavior of sodium alginate-g-poly (acrylic acid) hydrogel under DC electric field. *J. Macromol. Sci. A* 46: 1078–1082.

107 Yang, J.S., Xie, Y.J., and He, W. (2011). Research progress on chemical modification of alginate: a review. *Carbohydr. Polym.* 84: 33–39.

108 Yao, Q., Guo, J., and Wu, J. (2019). Preparation of chemically cross-linked modified sodium alginate/krill protein composite fiber. *J. Text. Res.* 40 (2): 8–13.

109 Yeom, C.K. and Lee, K.H. (1998). Characterization of sodium alginate membrane crosslinked with glutaraldehyde in pervaporation separation. *J. Appl. Polym. Sci.* 67: 209–219.

110 Zhan, Y., Zhao, X., Ma, J. et al. (2007). Preparation and properties of alginate/gelatin blend fibers. *Text. Aux.* 24 (8): 23–27.

111 Zhang, R., Wang, H., and Wen, B. (2004). *Preparation method for alginate viscose fiber*. Chinese Patent, CN1544727A

112 Zhang, S., Wang, H., Ni, F. et al. (2007). Clinical study on wound dressing of sustained-release alginate fiber. *Chin. Med. Equip. J.* 28 (10): 65–69.

113 Zhang, H., Gao, X., Guo, T. et al. (2011). Adsorption of iodide ions on a calcium alginate-silver chloride composite adsorbent. *Colloids Surf. A Physicochem. Eng. Asp.* 386: 166–171.

114 Zhang, J., Ji, Q., Tan, L. et al. (2011). Preparation and properties of nano-aluminum hydroxide/calcium alginate fibers. *J. Soc. Dye. Colour.* 5 (42): 820–822.

115 Zhu, H., Ji, J., Lin, R. et al. (2002). Surface engineering of poly(DL-lactic acid) by entrapment of alginate–amino acid derivatives for promotion of chondrogenesis. *Biomaterials* 23: 3141–3148.

10

Silver-Containing Alginate Fibers and Wound Dressings

10.1 Introduction

Infections in various forms have always been a major healthcare problem. Although many types of antibiotics have been developed over the years, the spread of antibiotic-resistant strains of microorganisms, such as methicillin-resistant *Staphylococcus aureus* (MRSA), represents an ever-increasing threat to the health of vulnerable people throughout the world who are obliged to spend extended periods in healthcare facilities. There is now a general agreement that the problem of resistance has been exacerbated by the overuse or misuse of antibiotics, and so wherever possible, alternative methods are required to manage topical infections.

Since they generally have a wet and warm condition that is a fertile ground for bacteria growth, open wounds act as an important source of cross-infection. In order to eliminate or prevent the spread of bacteria from such lesions, many types of wound dressings with antimicrobial properties have been developed and used over the years, often with the addition of various types of antimicrobial agents into the dressings. In recent years, the focus has been on the use of silver as a topical antimicrobial agent, and many wound dressings containing silver have been developed and launched in the wound management industry.

Silver has a long history as an antimicrobial agent, especially in the treatment of burns. While metallic silver is relatively inactive, silver ions are effective against a wide range of bacteria. When low concentrations of silver ions accumulate inside cells, they can bind to negatively charged components in proteins and nucleic acids, thereby affecting structural changes in bacterial cell walls, membranes, and nucleic acids that affect viability. Interestingly, although silver is a highly effective antimicrobial agent, it has limited toxicity to mammalian cells. It was found that in clean wounds in pigs, the use of silver-containing dressings can increase the rate of epithelialization by 28%, indicating a beneficial effect of silver ions in wounds, in addition to its antimicrobial activity.

As silver gained importance in the wound management industry as an effective antimicrobial agent, a large number of silver-containing wound dressings have been developed. These function by the sustained release of low concentrations of silver ions over time, and generally appear to stimulate healing, as well as inhibiting

Alginate Fibers and Wound Dressings: Seaweed Derived Natural Therapy, First Edition. Yimin Qin.

microorganisms. A number of laboratory studies have shown the excellent antimicrobial performances of silver-containing wound dressings. Human clinical studies have also shown encouraging clinical benefits.

Alginate wound dressings are an important type of modern wound management materials. Since they are highly absorbent, these dressings are mainly used on highly exuding wounds where microbial infection is common. By incorporating silver ions into alginate fibers, it is expected that a highly absorbent wound dressing with good antimicrobial properties can be obtained.

10.2 Antimicrobial Efficacy of Silver

The antimicrobial action of silver products has been directly related to the amount and rate of silver ions released and their ability to inactivate target bacterial and fungal cells [20]. The oligodynamic microbicidal action of silver compounds at low concentrations probably does not reflect any remarkable effect of a comparatively small number of ions on the cell, but rather the ability of bacteria, trypanosomes, and yeasts to take up and concentrate silver ions from very dilute solutions [4]. Therefore, bacteria killed by silver ions may contain 10^5–10^7 Ag^+ per cell, the same order of magnitude as the estimated number of enzyme–protein molecules per cell [5].

Chemically, metallic silver is relatively inert, but its interaction with moisture on the skin surface and with wound fluids leads to the release of silver ions and its biocidal properties. Silver ion is a highly reactive moiety and avidly binds to tissue proteins, causing structural changes in bacterial cell walls and intracellular and nuclear membranes [30]. These lead to cellular distortion and loss of viability. Silver binds to and denatures bacterial DNA and RNA, thereby inhibiting replication [29].

A study demonstrated the inhibitory action of silver on two strains of Gram-negative *Escherichia coli* and Gram-positive *S. aureus*. It found that silver-nitrate exposure leads to the formation of electron-light regions in their cytoplasm and the condensation of DNA molecules [12]. Granules of silver were observed in the cytoplasm, but RNA and DNA damage and protein inactivation seemed to be the principal mechanisms for bacteriostasis. Silver-related degenerative changes in bacterial RNA and DNA, mitochondrial respiration, and cytosolic protein lead to cell death.

The action of silver ion on cell walls is illustrated by reference to the yeast *Candida albicans*. Silver has been shown to inhibit the enzyme phosphomannose isomerase (PIM) by binding cysteine residues [38]. This enzyme plays an essential role in the synthesis of the yeast cell wall, and defects lead to the release of phosphate, glutamine, and other vital nutrients.

Literature studies suggest that the microbicidal action of silver products is partly related to the inhibitory action of silver ions on cellular respiration and cellular function [8]. The exact nature of these silver radicals is not clear, but Ovington [30] noted that nanocrystalline silver products (Acticoat, Smith & Nephew) can release a cluster of highly reactive silver cations and radicals, which provide a high antibacterial potency on account of unpaired electrons in outer orbitals. Silver

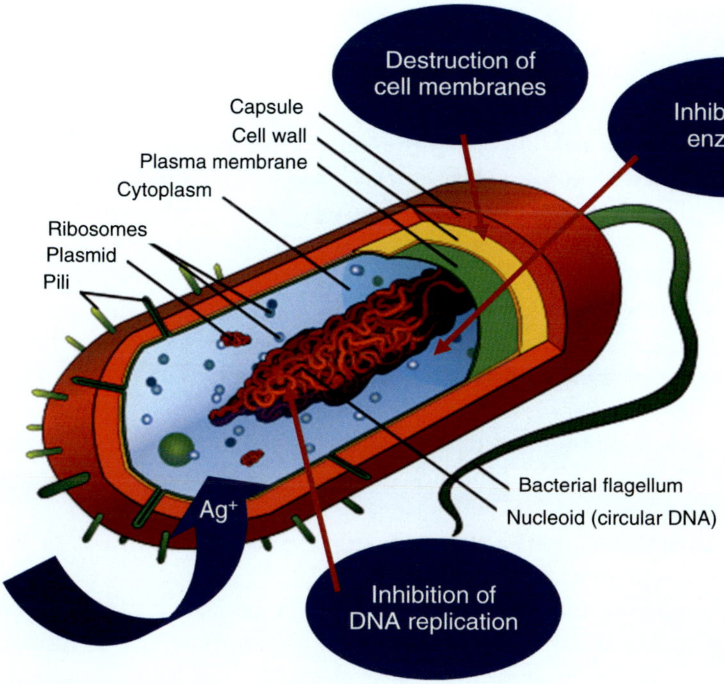

Figure 10.1 The antimicrobial mechanism of silver ion.

and silver radicals released from Acticoat also cause impaired electron transport, bacterial DNA inactivation, cell membrane damage, and binding and precipitation of insoluble complexes with cytosolic anions, proteins, and sulfhydryl moieties. Figure 10.1 summarizes the main antimicrobial mechanism of silver ions.

10.3 Development of Silver-Containing Wound Dressings

Historically, silver is more than a decorative metal. Romans used silver containers to prevent milk from deteriorating. For centuries, physicians in the field of wound management have exploited the curative properties of silver to assist in the healing of burns and chronic wounds. Surgeons used silver regularly in the treatment of open abscesses, leg ulcers, and venereal swelling. With regard to the antimicrobial properties of silver, as early as 1893, Naegeli had reported that a silver concentration of 0.0000001% killed the freshwater algae *Spirogyra* and a concentration of 0.00006% prevented germination of *Aspergillus niger* spores [33]. However, with the advent of antibiotics such as penicillin and sulfonamides, the nineteenth century saw a decline in the use of silver. In recent decades, as antibiotic resistance has become common, silver has been revived to treat wounds, including venous leg ulcers, diabetic foot ulcers, and burn wounds. In the 1960s, the late Dr. Carl Moyer, chairman of the Washington University School of Surgery, explored silver's role in burn therapy. Despite antibiotic use, he noticed that infection of burns remained a serious problem. He used silver nitrate solutions, silver sulfadiazine (SSD) cream, and dressings containing elemental or ionized silver that is applied to thick cotton gauze dressings to heal burns as well as grafted skin. Dr. Moyer noticed the approach was effective against several types of infection, and his work spawned many of the advances that have been seen in the area of silver-based wound management products.

The excellent biocidal properties of silver led to the development of many silver-containing wound dressings. Deitch [7] evaluated the antimicrobial activity of two silver–nylon fabrics and showed them to be microbicidal *in vitro* against *S. aureus*, *Pseudomonas aeruginosa* and *C. albicans*. Ersek et al. [11] described how silver impregnated into aldehyde cross-linked skin or a porcine xenograft could be used to decontaminate and promote healing of massive and chronically contaminated longstanding wounds. In recent years, silver has been further shown to be active against bacterial, fungal, and viral pathogens. It can be applied topically in cream form or as a wound dressing to treat wounds. In cream and dressing form, the silver remains active within the wound for several days, thus avoiding the toxicity that results from immediate absorption.

The first major silver-containing dressing to make a commercial impact was Actisorb Plus (now Actisorb Silver 220, Johnson & Johnson), which is primarily composed of a silver-impregnated activated charcoal cloth, as illustrated in Figure 10.2. After comparing the antimicrobial properties of the silver-containing formulation with those of the original unmedicated product, Actisorb, using a zone of inhibition technique [13], it was found that Actisorb Plus inhibited the growth

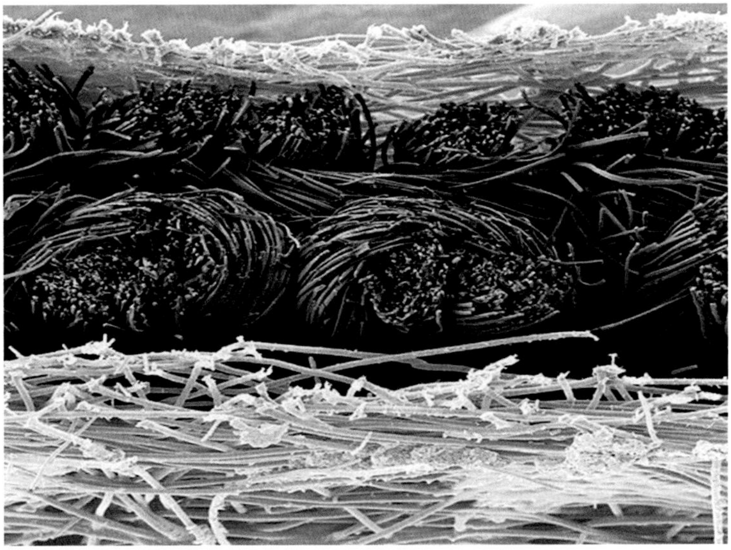

Figure 10.2 Cross-sectional structure of Actisorb Silver 220 from Johnson & Johnson, showing silver-containing activated carbon fabric embedded in two layers of polyamide nonwoven fabric.

of a range of different bacteria. Further tests showed that this antimicrobial action was due to the release of low concentrations of silver ions from the dressing.

In 1998, Tredget et al. [36] reported the development of Acticoat (a silver-coated high-density polyethylene membrane now supplied by Smith & Nephew) and described a randomized prospective clinical study involving 30 patients, each of whom had two burns comparable in size, depth, and location. The wounds in each pair were treated with Acticoat or a fine mesh gauze soaked in 0.5% silver nitrate solution and re-moistened every two hours. It was found that the frequency of burn wound sepsis ($>10^5$ organisms per gram of tissue) was less in Acticoat-treated wounds than in those treated with silver nitrate (5 vs 16). Secondary bacteremias arising from infected wounds were less frequent with Acticoat (1 vs 5).

After many years of development, silver in various forms has been incorporated into a wide range of wound dressings made of different materials, such as hydrogel, alginate, and polyurethane foam. These silver-containing dressings are now commonly used on burns, skin graft and donor sites, surgical wounds, diabetic foot ulcers, leg ulcers, pressure ulcers, and other wounds where clinical infection can be inhibited.

10.4 Applications of Silver in Alginate Fibers and Wound Dressings

10.4.1 Types of Silver Compounds Used in Wound Dressings

Silver is a group 11 element (formerly group Ib) of the periodic table and exists as two isotopes, ^{107}Ag and ^{109}Ag, in approximately equal proportions. In solution, silver exhibits three oxidation states, i.e. Ag^{+}, Ag^{++}, and Ag^{+++}, each capable of forming inorganic and organic compounds and chemical complexes. Compounds involving Ag^{++} or Ag^{+++} are unstable or insoluble in water.

The silver compounds used in wound dressings can be divided into the following three groups:

1. Elemental silver, e.g. nanocrystalline particles or foil;
2. Inorganic compounds/complexes, e.g. silver nitrate, silver sulphadiazine, silver oxide, silver phosphate, silver chloride, or a silver zirconium compound;
3. Organic complexes, e.g. colloidal silver preparations, silver-zinc allantoinate, or silver proteins.

Colloidal silver solutions were the most common delivery system prior to 1960. It is in the form of charged pure silver particles (3–5 ppm) held in suspension by small electric current, where the positively charged ions repel each, hence making it possible for the particles to remain in solution when applied topically to a wound. Although it is highly bacteriocidal with no resistance, since the solutions are unstable when exposed to light, colloidal silver offers little practical value. When the silver ions are complexed into small proteins to improve stability in solution, they become more stable. However, they are also much less antibacterial than pure ionic silver.

In the 1960s, various silver salts were developed. When silver ions are complexed to AgCl, $AgNO_3$, and Ag_2SO_4, they become more stable delivery systems. Silver nitrate was the most widely used compound, but it is dangerous to use in concentrations exceeding 2%. An aqueous solution containing 0.5% silver nitrate is the standard solution for treating burns and infected wounds. However, nitrate is toxic to wounds and cells and appears to decrease healing. It is also unstable in light.

Various studies have shown that pure silver ions and radicals produce the best antimicrobial results, as well as optimize the wound healing environment. As a consequence, the silver salts and complexes used today were developed to maintain a sustained release of silver ions. A typical silver-containing compound is Alphasan RC5000 developed by Milliken. As a zirconium phosphate-based ceramic ion-exchange resin containing silver, Alphasan is effective against a range of microorganisms that can cause undesirable effects. The material is widely used in Europe, Japan, and the United States, and it has been approved by US FDA for contact applications.

Table 10.1 summarizes various silver compounds used in different silver-containing wound dressings.

10.4.2 Methods for Adding Silver to Wound Dressings

The silver-containing wound dressings currently available have considerable differences between their overall structure, the concentration, and formulation of the silver compounds. Overall, silver ions can be added to wound dressings using the following four basic methods [31]:

- Chemical treatment of base material. In this method, the fiber or fabric can be coated with metallic silver or treated with silver-containing solutions, whereby silver ions can be attached to the wound dressing through ion exchange;
- Physical treatment of the base material. In this method, the fiber or fabric can be coated with metallic silver;
- Blending. Fine particles of the silver compounds can be blended with the base material;

Table 10.1 Silver compounds used in various wound dressings.

Manufacturer	Name of the product	Silver compound used
Argentum	Silverlon	Metallic silver
Smith & Nephew	Acticoat	Metallic silver
Medline	SilvaSorb	Silver chloride
Convatec	Aquacel Ag	Silver chloride
Medline	Arglaes	Silver calcium phosphate
Coloplast	Contreet	Silver ammonium complex
Johnson & Johnson	Actisorb Silver 220	Silver carbon

- Blending of silver-containing fibers with other fibers. This method is used in the production of Silvercel, where alginate fibers are blended with the silver-coated X-static fibers.

10.4.3 Examples of Silver-Containing Wound Dressings

A brief description of some of the silver-containing wound dressings is given below.

10.4.3.1 Acticoat from Smith & Nephew

Acticoat consists of two layers of a silver-coated, high-density polyethylene mesh, enclosing a single layer of an apertured nonwoven rayon and polyester fabric. These three components are ultrasonically welded together to maintain the integrity of the dressing while in use. Silver is applied to the polyethylene mesh by a vapor deposition process, which results in the formation of microscopic crystals of metallic silver. Upon activation with water, Acticoat provides a rapid and sustained release of silver ions within the dressing and to the wound bed for three or seven days.

10.4.3.2 Silvercel from Johnson & Johnson

Silvercel combines the potent broad-spectrum antimicrobial action of silver with enhanced exudate management properties of alginate technology. Because of a sustained release of silver ions, the dressing acts as an effective barrier and helps reduce infection. As is shown in Figure 10.3, the antimicrobial properties are built in through the use of X-Static silver-coated fibers blended into the nonwoven structure. Silvercel wound dressing has been proven effective *in vitro* against 150 clinically isolated microorganisms, including antibiotic-resistant strains.

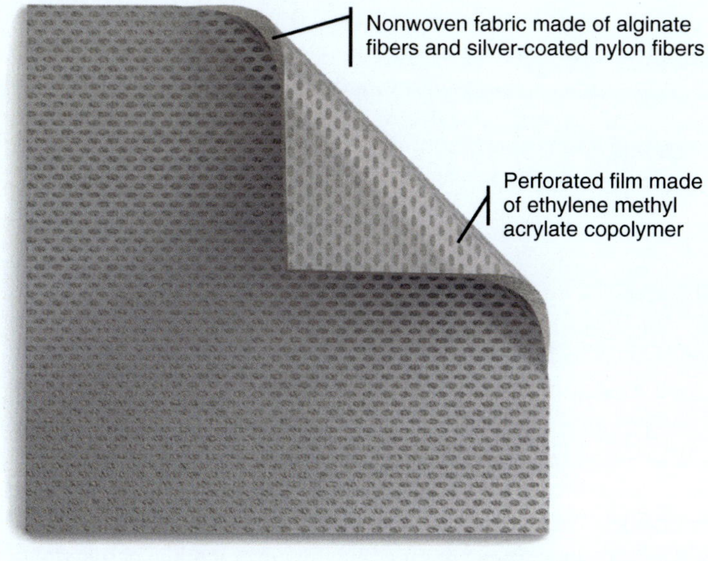

Figure 10.3 An illustration of Silvercel wound dressing.

10.4.3.3 Aquacel Ag from ConvaTec

Aquacel consists of a fleece of sodium carboxymethylcellulose fibers containing 1.2% ionic silver. In the presence of exudate, the dressing absorbs liquid into the fiber structure to form a fibrous gel, and at the same time, releases silver ions to effect antimicrobial actions.

10.4.3.4 Contreet Foam from Coloplast

Contreet Foam (Coloplast A/S) comprises a soft hydrophilic polyurethane foam containing silver as an integral part of the matrix. It is recommended for use on moderately to highly exuding chronic wounds with a high bacterial burden. Silver ions are present in a form that is readily "hydroactivated" in the presence of fluid or wound exudate.

10.4.3.5 Silverlon from Argentum Medical

Silverlon is a knitted fabric dressing that has been silver-plated by means of a proprietary autocatalytic electroless chemical (reduction–oxidation) plating technique. This technique coats the entire surface of each individual fiber from which the dressing is made, resulting in a very large surface area for the release of ionic silver.

10.4.3.6 SilvaSorb from Medline

SilvaSorb Silver Antimicrobial Wound Gel is an amorphous gel wound dressing for use in moist wound care management. Its SilvaSorb MicroLattice® technology maintains an optimally moist wound environment by either absorbing slight drainage or donating moisture while it delivers antimicrobial ionic silver. SilvaSorb Silver Antimicrobial Wound Gel can act as an effective antimicrobial barrier.

10.4.3.7 Urgotul SSD from Laboratoires URGO

Urgotul SSD dressing comprises a polyester mesh impregnated with carboxymethylcellulose, vaseline, and silver sulphadiazine (3.75%). SSD is composed of sulfonamide, which is bacteriostatic, and silver, which is bactericidal. Its mechanism of action results from the synergetic activity of the sulfonamide and silver components, which inhibit the replication of bacterial DNA [28].

10.4.3.8 Actisorb Silver 220 from Johnson & Johnson

Actisorb silver 220 consists principally of activated carbon impregnated with metallic silver, produced by heating a specially treated fine viscose fabric under carefully controlled conditions. The carbonized fabric is enclosed in a sleeve of spun-bonded nonwoven nylon, sealed along all four edges, to facilitate handling and reduce particle and fiber loss. When applied to a wound, the dressing adsorbs toxins and wound degradation products, as well as volatile amines and fatty acids responsible for the production of wound odor. Bacteria present in wound exudate are also attracted to the surface of the dressing where they are killed by the antimicrobial activity of the silver, which is active against a wide range of pathogenic organisms.

Table 10.2 Typical silver contents of the silver-containing wound dressings.

Product name	Ag content (mg/100 cm^2)
Silverlon	546
Calgitrol Ag	141
Acticoat	105
Contreet Foam	85
Contreet Hydrocolloid	32
Aquacel Ag	8.3
SilvaSorb	5.3
Actisorb Silver 220	2.7
Arglaes powder	6.87 mg/g

10.4.3.9 Microbisan from Lendell Manufacturing Inc.

Microbisan absorbent dressing was jointly developed by Lendell Manufacturing Inc. and Milliken Inc. In this product, the antimicrobial silver particles are added to the foam by means of a polyurethane production process that completely disperses the additive throughout the finished product. The antimicrobial agent in the foam keeps the wound free of bacterial contamination as the dressing absorbs wound fluids.

Because of the differences in the types of silver compounds and the techniques in applying them to the wound dressings, different silver-containing wound dressings have considerably different silver contents [23, 34, 35]. Table 10.2 shows the typical silver contents of the silver-containing wound dressings.

10.4.4 Differences Between Silver-Containing Wound Dressings

As can be seen in the previous section, the various types of commercially available silver-containing wound dressings differ in their base materials and in the silver compounds used for delivering silver ions. These differences are briefly analyzed below.

10.4.4.1 Different Silver Compounds

Various types of silver compounds are used in the manufacture of silver-containing wound dressings. In order to achieve sustained resistance to microbial infection, the ideal silver-containing wound dressing should be able to release silver ions over the duration of the dressing's useful period. In this respect, the commonly used 0.5% silver nitrate solution releases all the silver ions on application, and most of the available silver ions are then combined with sodium chloride or protein to form insoluble silver chloride and silver sulfide. Silver nitrate solution had to be frequently re-applied while argyria is also easy to form.

Although SSD has been widely used in burn wound management, it has been found that the sulfadiazine component poses a certain level of toxicity to human body and bacteria resistance has also been observed for this compound [15].

Metallic silver has a limited solubility in water, and when it is exposed to wound fluid, only a small amount of silver ions are released. When metallic silver is coated onto a textile substrate to act as the silver reservoir, it has to be applied in a large amount and over an extended area to deliver an effective dose of silver during application.

Silver-containing chemical complexes are the most commonly used silver reservoir. Taking the Alphasan compound as an example, since they have very fine particle sizes, they can be evenly distributed into the fiber or foam structures, hence they have a large area of exposure to the wound fluid, which can activate the release of silver ions. These compounds help to protect the base material from being oxidized, while at the same time, can maintain a sustained release of silver ions during application.

10.4.4.2 Different Contact Areas

For the silver ions to release into the wound environment, the silver-containing reservoirs in various forms have to be exposed to the wound fluid. If the silver reservoir is embedded inside the fiber or foam structure, then the release of silver ions is slow and incomplete. On the other hand, if the silver is coated on the surface of the wound dressing, the release of silver ions can be fast. Both the Silverlon and Acticoat products are coated with metallic silver. The Silverlon product is a knitted fabric and silver is coated onto the surface of individual fibers, while in Acticoat, silver is coated on the surface of a plastic membrane. Since the contact area for Silverlon is about 50 times that of Acticoat, the release of silver from Silverlon is also faster than from Acticoat.

10.4.4.3 Different Absorption Capacities

In addition to the antimicrobial properties, silver-containing wound dressings are also used to absorb exudate from the wound, hence absorption capacities are also important. Figure 10.4 shows the absorption capacities of several silver-containing

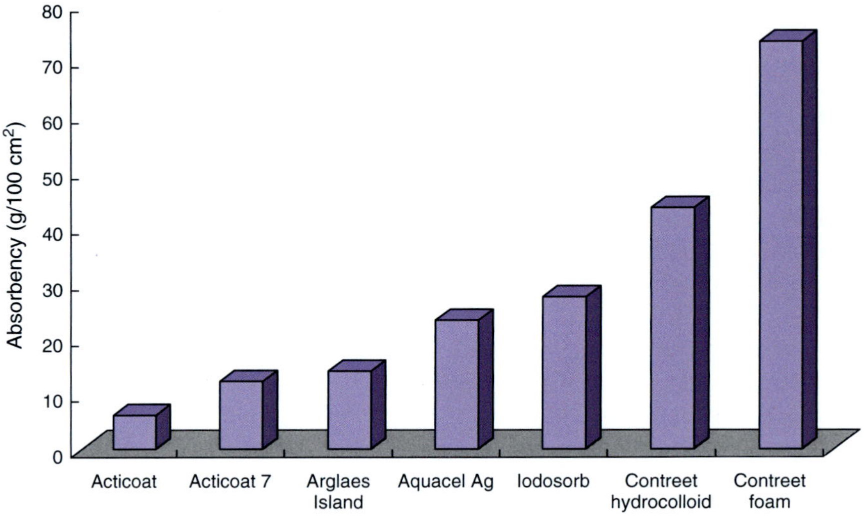

Figure 10.4 Absorption capacities of various silver-containing wound dressings.

wound dressings. It can be seen that because the base materials are different, the absorption capacities are significantly different from each other. Since Acticoat used plastic film as a carrier for silver, it has a low-absorption capacity. On the other hand, Contreet Foam used a hydrophilic polyurethane foam as the base material, and because the foam has a highly porous structure, its absorption capacity is the highest among the silver-containing wound dressings.

10.5 Preparation of Silver-Containing Alginate Fibers and Wound Dressings

10.5.1 The Addition of Silver Into Alginate Fibers Through Chemical Reaction

Since alginate is a polymeric acid, it can form salt with silver ions. However, when sodium alginate solution is extruded into a silver nitrate solution, it is difficult to form silver alginate fiber, mainly because silver is a monovalent ion and solidification of the newly spun filament is slow and difficult. A mixed solution of calcium chloride and silver nitrate can be used to produce fibers that are a mixture of calcium alginate and silver alginate. In another method, calcium alginate fibers can be treated with aqueous solutions of silver nitrate. The silver ions in the solution exchange with the calcium ions in the fiber, resulting in the formation of calcium alginate fiber-containing silver ions.

10.5.2 The Addition of Silver Into Alginate Fibers Through Blending

Although silver is a highly effective broad-spectrum antimicrobial agent, it is also highly oxidative to organic materials. Skin discoloration and irritation associated with the use of silver nitrate is well known. In order to protect the host material from oxidation and discoloration, some novel silver-containing compounds have been developed and have been made into fine particles that can be blended with fiber-forming polymers during extrusion.

Alphasan RC5000 is a silver sodium hydrogen zirconium phosphate. This microbiologically active ingredient is a synthetic inorganic polymer that resembles cube-shaped crystals, with an average particle size of about 1 μm (about the size of an average bacterium). It consists of a three-dimensional, repeating framework of sodium hydrogen zirconium phosphate, with many equally spaced cavities containing silver. Silver (at 3.8% by weight) provides the main anti-microbial properties, while the framework matrix acts to distribute silver evenly (without clumping or pooling) throughout the individual fibers where the Alphasan particles are added.

When Alphasan RC5000 is mixed with sodium alginate solution, the fine particles can be evenly distributed in the spinning solution under a high rate of shearing. Because the particles are very fine, they can be suspended uniformly while the solution is extruded to form fibers. Since the sodium hydrogen zirconium phosphate framework prevents the silver ions from oxidizing the alginate, this type of silver-containing alginate fiber remains white even after sterilization through irradiation. Figure 10.5 shows the photomicrograph of alginate fiber with Alphasan RC5000 particles uniformly distributed inside the fiber structure.

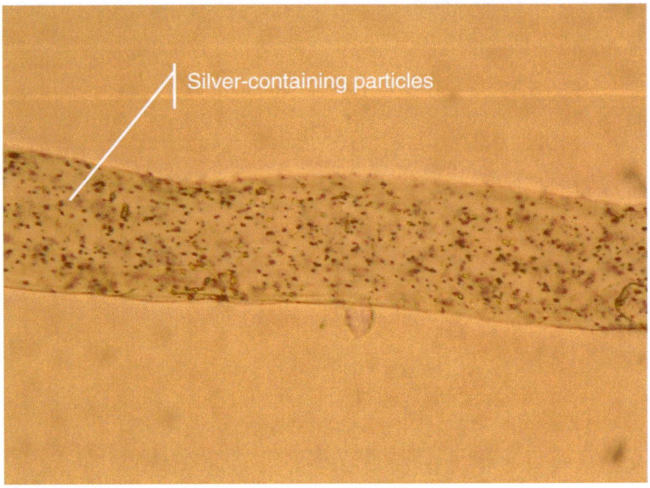

Figure 10.5 Photomicrograph of alginate fiber with Alphasan RC5000 particles uniformly distributed inside the fiber structure, X200. Adapted with permission from Figure 11.1 in Qin [31].

10.6 Release of Silver Ions from Silver-Containing Alginate Fibers

When alginate fibers containing Alphasan RC5000 particles are in contact with wound exudate, the silver ions can be released into the wound exudate by three mechanisms. First, there is an ion exchange between the silver ions in the fiber and the sodium and calcium ions in the wound fluid. Second, silver ions can be chelated by protein molecules in the wound fluid. Third, Alphasan particles attached on the surface of the fibers can also be detached from the fibers and get into the wound exudate.

Table 10.3 Silver concentrations in the contact solutions.

Sample	Duration of contact	Silver concentration in the contact solution (mg/l)
	The following uses normal saline	
1	30 min	0.50
2	48 h	0.40
3	7 days	1.32
	The following uses human serum	
A	30 min	2.18
B	48 h	2.74
C	7 days	3.74

Table 10.3 shows the silver ion concentration when alginate fiber containing 1% Alphasan RC5000 is placed in contact with normal saline or human serum. It can be seen that the silver ions are slowly released into the solution, acting as an antimicrobial agent. More silver ions can be seen to release into human serum, suggesting the high silver-binding abilities of the protein components in the wound exudates.

10.7 The Antimicrobial Effect of Silver-Containing Alginate Fibers and Wound Dressings

It is well known that when they are in contact with wound exudate, the calcium ions in the calcium alginate fibers exchange with sodium ions in the fluid, and the fibers are transformed from water-insoluble calcium alginate into water-soluble sodium alginate, resulting in the absorption of a large amount of water by the fibers. In an alginate wound dressing, typically with a nonwoven structure, as the fibers absorb water and swell, the spaces between the fibers are closed and any bacteria that is carried in the wound exudate is trapped in the wound dressing. This can help to reduce the spreading of bacteria, giving the alginate wound dressings bacteria static properties.

The silver ions in the alginate fibers can enhance their antimicrobial activities by the sustained release of the broad spectrum antimicrobial silver ions. As Table 10.3 shows, silver ions can be released over a 7-day period. These ions can kill the bacteria that are trapped in the alginate wound dressing, thus making the dressings bacteriocidal. Figure 10.6 shows the antimicrobial mechanism of the silver-containing alginate wound dressings.

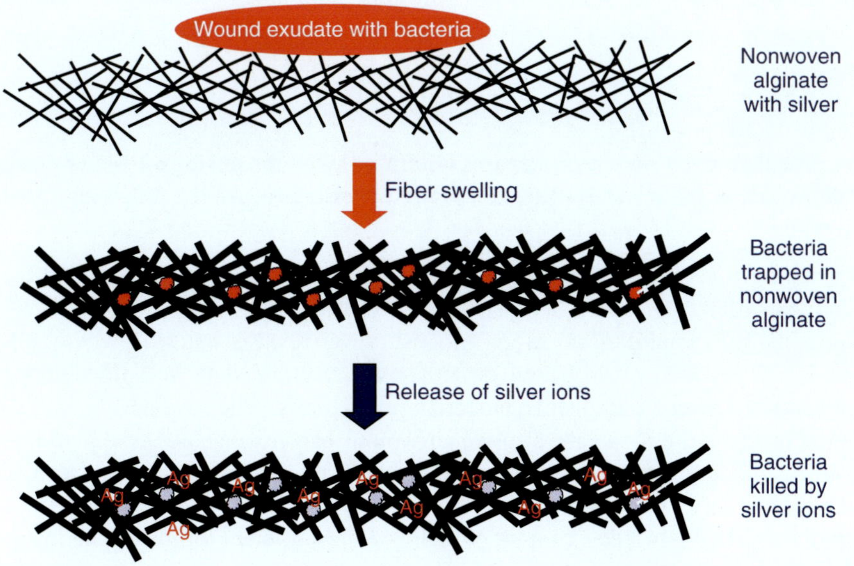

Figure 10.6 Antimicrobial mechanism of the silver-containing alginate wound dressings.

10.8 Properties and Applications of Silver-Containing Alginate Wound Dressings

Since ionic silver exhibits antimicrobial activity against a broad range of microorganisms, silver is now included in many commercially available healthcare products, and the use of silver is now common in the field of wound management. A wide variety of silver-containing wound dressings are now available.

10.8.1 Wound Healing Properties of Silver

The anti-inflammatory effects of silver ions on wounds have been recognized for centuries. Most of the reports are purely descriptive in nature, identifying the decrease in erythema and accelerated healing. Although a number of the biochemical effects of silver on the wound have been documented, only recently with the new concepts on wound healing and healing impairment, can a mechanism of action be presented. Studies suggest that the silver-coated Acticoat wound dressing influences wound healing through its action on matrix metalloproteinases [19, 41], although its mechanism of action is not fully understood at present time. It is suggested that the silver ion released suppresses neutrophil influx through inflammatory cytokines (IL-1 and TNF-α). These facultative changes might reflect a "compression" of the inflammatory phase, hence reducing the overall time for the wound to heal.

10.8.2 The Release of Silver from Silver-Containing Wound Dressings

Sustained silver-releasing dressings are now well established in the management of chronic wounds and burns [21, 32]. The dressings differ greatly in their total silver content, patterns of silver ion release, and presumed therapeutic action [22]. During application, wound exudate triggers the release of "activated" silver ions for antibacterial action and to neutralize any toxins. It is estimated that in order to achieve effective bactericidal action, the silver concentration in tissue culture should ideally be about 10–40 ppm [14].

In a detailed study on various silver-containing wound dressings, Dolmer et al. [10] showed that the release profiles for wound dressings are of the following three main types:

1. Products with a high silver content with a rapid silver ion release designed for wounds with heavy exudate and bacterial colonization;
2. Products that maintain a more modest silver-release pattern, where silver ion is released over several days. These are claimed to be sufficient for moderate-to-severe pathogenic bacterial populations. The non-silver components of these dressings are attuned to wound bed management, i.e. exudate control, debridement of wound debris, and management of the wound environment;
3. Products with a low silver content, which may be sufficient for low-grade infections in chronic wounds but are more appropriately used as a barrier to infection in acute wounds, burns, and surgical injuries.

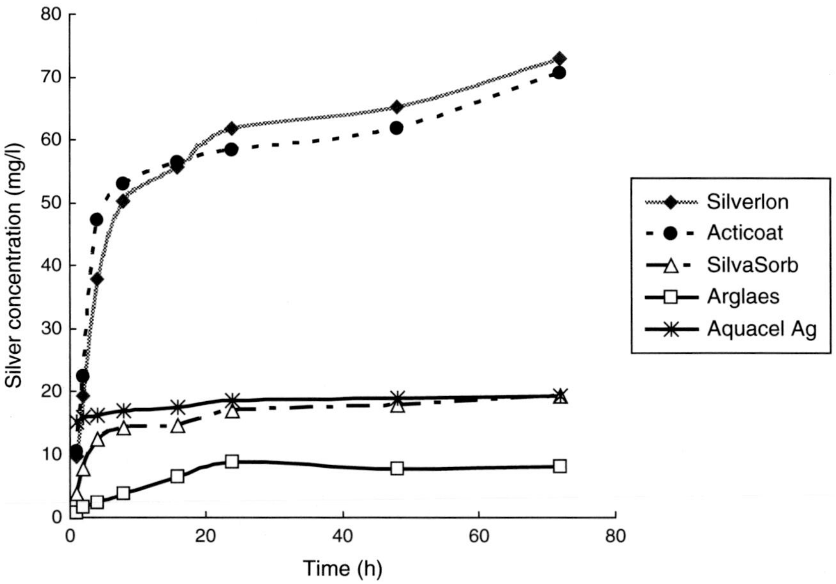

Figure 10.7 Silver concentration in the contact solution after the silver-containing dressings are placed in contact with wound fluid over different periods of time. Adapted with permission from Figure 11.3 in Qin [31].

Figure 10.7 shows the silver concentration in the contact solution after the silver-containing dressings are placed in contact with wound fluid over different periods of time. It can be seen that Silverlon and Acticoat release more silver ions than other types of products. Both the Silverlon and Acticoat products are coated with a high level of metallic silver. A high level of silver release is needed since both products are intended for burn wounds where the prevention of infection is an important consideration. In other types of dressings, the silver ions act as an antimicrobial agent to control bacteria growth and prevent cross-infection. With silver-containing alginate wound dressings, the dressing first absorbs exudate from wound bed into the dressing structure. As the dressing becomes wet, silver ions are activated and liberated into the contacting solution, exerting antimicrobial effect on the bacteria. In these cases, the level of silver release can be significantly lower than in burn wound management.

Lansdown et al. [26] made a sequential microbiological examination of wound swabs, sampling of wound exudate, and wound scale from seven patients with chronic wounds. After analyzing the silver content using atomic absorption spectrometry, they found that:

- All silver released into the wound bed was absorbed by wound exudate or debris (wound scale, etc.);
- Silver uptake by wound exudate is approximately proportional to its viscosity (protein content);
- Silver absorbed into the wound bed may be eliminated in the exudate for several weeks following the termination of silver therapy;

- The amount of silver released from dressings is closely related to the amount of moisture absorbed;
- Wounds treated with silver do not attain a germ-free status, suggesting that silver-resistant organisms such as *S. aureus* and *P. aeruginosa* may contribute to indolence in healing.

10.9 Test Methods for Assessing the Antimicrobial Properties of the Silver Dressing

Silver-containing wound dressings are used mainly for the treatment of wounds with infection. Three test methods have been designed to compare the performance of dressings under different simulated conditions of use [34, 35]. The zone of inhibition method simulates the use of the products on moist or lightly exuding wounds and predicts the dressings' ability to kill or prevent bacterial growth in this situation. In order to exert a significant antimicrobial effect in this test, a dressing must first absorb moisture to activate or release the silver held within its structure, which exerts its antimicrobial action. The microbiological challenge test provides an indication of each dressing's ability to kill or prevent growth of predetermined numbers of bacteria applied directly in the form of a suspension, and thus to some extent reflects what may occur within dressings applied to more heavily exuding wounds. The microbial transmission test determines the bacteria's ability to survive and be transmitted along the dressing surface.

10.9.1 Zone of Inhibition

Petri dishes containing a 5-mm layer of tryptone soya agar for bacteria and Sabouraud agar for the yeast were inoculated with 0.2 ml of a log-phase broth culture of each test organism. This suspension was distributed uniformly over the surface of the plate and allowed to dry for 15 minutes. Portions of each dressing measuring 40 mm × 40 mm were then placed on the agar, with the wound contact surface downwards. The plates containing the bacteria were incubated for 24 hours at 35 °C, and the plates with the fungi at 20–25 °C for 48 hours. At the end of the incubation period, the plates were examined for microbial growth and the presence of a zone of inhibition. If detected, the width was measured and the dressing was removed from the agar and placed on another agar plate, seeded as before with the same microorganism. This process was repeated a maximum of five times or until no further zone of inhibition was produced.

As can be seen in Figure 10.8, the zone of inhibition method can be simplified by placing a small sample in a bacteria suspension. The antimicrobial effect of the product can be observed by the clarity of the solution around the sample.

10.9.2 Challenge Testing

A total of 0.2 ml of a log- phase culture from each microorganism was added to portions of each dressing measuring 40 mm × 40 mm. The inoculated dressings

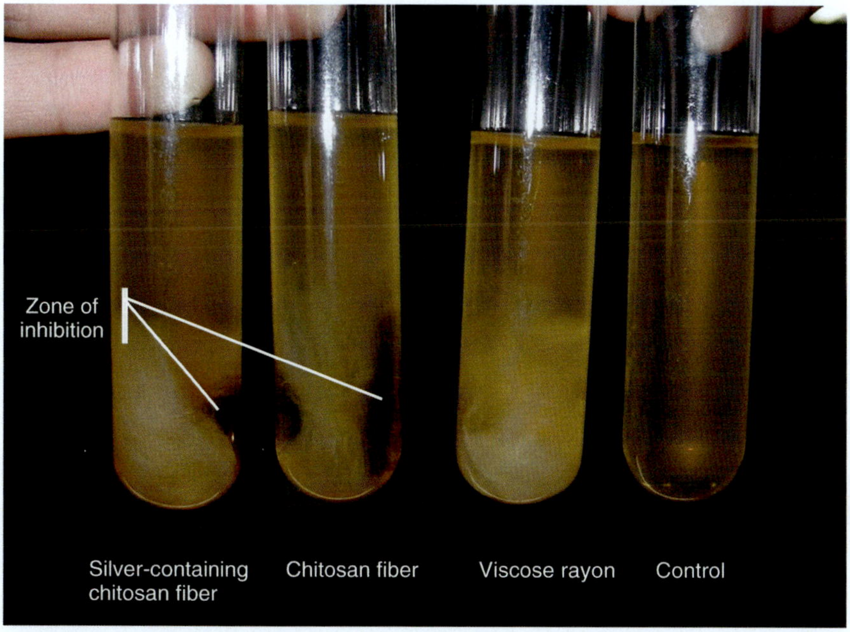

Figure 10.8 Clarity of solutions in the control sample and in suspension of *Escherichia coli* with silver-containing chitosan fiber, chitosan fiber, and viscose rayon.

were incubated for two hours, then transferred into 10 ml of 0.1% peptone water (Oxoid) and vortexed to remove any viable organisms remaining in the dressings. Serial dilutions were performed in triplicate on each extract, and the number of viable organisms present in each was determined using a standard surface counting technique. If viable organisms were recovered during this process, the test was repeated using a 4-h contact period, and then again with a 24-h contact period. If no organisms were detected after two hours, the dressing was placed in 10 ml tryptone soya broth (TSB) to detect very low levels of residual contamination, effectively a form of sterility test.

10.9.3 Microbial Transmission Test

A strip of agar was removed from a standard agar plate, leaving two areas of agar separated by a narrow channel. A second strip was removed on the outer side of one of these areas, effectively forming one of the agar areas into an island. This island was inoculated with a bacterial suspension as before. After a short drying period, three strips of the dressings under test, approximately 10 mm wide and 50 mm long, were placed across the two agar areas forming a bridge. The fluid reservoir on the outer side of the agar island was then filled with sterile water, and the plate was incubated in the upright position for 24 hours. During this period, the agar absorbed some of the added water. This was taken up by the dressing strip and distributed throughout its length, carrying with it bacteria drawn up from the surface of the agar island. In the absence of significant antimicrobial activity within the dressing, these microorganisms would eventually reach the second area of agar, where they would grow to form colonies around the margin of the test sample.

10.10 *In Vitro* and *In Vivo* Findings of the Clinical Benefits of Silver in Wound Healing

In an in vitro study, Wright et al. [39] compared the antimicrobial properties of Acticoat with a solution of silver nitrate and cream containing SSD against 11 antibiotic multi-resistant clinical isolates. Acticoat was most effective at killing the organisms. A later study compared the activity of the same dressings against a spectrum of common burn wound fungal pathogens and showed that the silver-coated membrane provided the fastest and broadest-spectrum fungicidal activity [40].

Yin et al. [42] compared the antimicrobial activity of Acticoat with silver nitrate, silver sulphadiazine, and mafenide acetate in order to determine their minimum inhibitory concentration (MIC), minimum bactericidal concentration (MBC), and zone of inhibition. Although mafenide acetate produced the greatest zone of inhibition, the MBC of the product was higher than its MIC, indicating that it had a bacteriostatic rather than a bactericidal action. In contrast, the MICs and MBCs of the silver-containing products were similar, indicating that their activity is essentially bactericidal. The authors showed that, although the MIC values for the three silver preparations were very similar when calculated in terms of their silver content, Acticoat acted more rapidly than the other two products, perhaps because the metallic silver on the surface of the dressing forms a reservoir of silver ions, which are released continuously and are, therefore, always available for bacterial uptake.

The results of the microbiological tests demonstrate that major differences exist in the properties of the different silver-containing dressings. Not unexpectedly, these appear to reflect the silver content of the products concerned, as determined by chemical analysis, which varies from 1.6 to 109 $mg/100\,cm^2$. In a detailed study on the performances of various silver-containing wound dressings, Thomas et al. [34, 35] found that when testing against *S. aureus:*

- Acticoat exhibited a marked bactericidal effect within two hours;
- Contreet-H had an inhibitory effect;
- Actisorb Silver 220 appeared to prevent the proliferation of the organism within the dressing after a minimum contact time of four hours. Actisorb Silver 220 removed the organisms from the suspension and bound them to the surface of the charcoal fibers. These organisms remain viable for many hours until they are progressively inactivated by the silver ions in the dressing. The inner core, which contains silver, forms an effective barrier, whereas the outer nylon sleeve does not.

While total silver content is important, other factors also influence a dressing's ability to kill microorganisms. These include the distribution of the silver within the dressing (whether it is present as a surface coating or is dispersed through the structure), its chemical and physical form (whether it is present in a metallic, bound, or ionic state), and the dressing's affinity for moisture, a prerequisite for the release of active agents in an aqueous environment. Products in which the silver content is concentrated on the dressing surface rather than "locked up" within its structure performed well, as did those in which silver was present in the ionic form.

Argyria (argyrosis) is probably the most common contra-indication of using silver compounds in wound dressings and medicaments. Argyria may arise through the application of silver medication to open wounds, where silver released from soluble salts in the presence of light leads to the precipitation of black silver sulfide. In the skin, the black granules of silver metal or silver sulfide (estimated to be 30–100 nm in size) are most common around sweat and sebaceous glands, hair follicles, and nail bed and in the region immediately below the basement membrane [1, 3]. Substantial clinical and experimental evidence shows that although cosmetically undesirable in appearance, the silver deposits in skin are transient and can be eliminated during the normal course of wound healing. There is no evidence that they present a health risk [27].

10.11 Local and Systemic Toxicity of Silver in Wound Healing

Silver is commonly present in the human body, albeit at exceedingly low concentrations. When silver is absorbed into the body, some of it precipitates as silver sulfide or silver chloride in the presence of sodium chloride in sweat, or binds in a stable complex with proteins at the wound site and in exudate. In due course, some silver is eliminated through the skin, with the balance being voided in the urine and feces.

Silver poses minimal health risks to humans through occupational or therapeutic exposure, and clinicians widely accept that it is a highly effective antibacterial agent in wound care [16]. Absorption of silver by mammalian tissue is not well documented, but biochemical evidence indicates that it avidly binds to proteins, including those present in the wound site, and can influence trace metal metabolism [24]. Silver is not a trace metal itself, and like other foreign substances, it exerts toxic changes under some circumstances. To be toxic, silver needs to be absorbed into the body in sufficient amounts to evoke irreversible changes in a target organ.

It has been reported that absorption of ionized silver through intact skin is low in all parts of the body [17]. Less than 4% was absorbed by patients treated with aqueous solutions or creams containing 0.5–2% silver nitrate over 24–48 hours. No measurable changes in serum silver (normal level: $<2.3\ \mu g/l$) were noted in two volunteers exposed daily to topical SSD for two weeks [6, 37]. However, workers occupationally exposed to silver have shown serum levels of $11\ \mu g/l$ [9]. Absorption of silver through damaged or wounded skin is appreciably higher than through intact skin, and might be related to the following factors:

- the depth and extent of the wound;
- the mode and frequency of application;

- the silver content of the therapeutic or dressing;
- the amount of exudate, which influences the amount of ionized silver released.

In a study of 22 patients with burn injuries, Coombs et al. [6] treated the wounds with silver sulphadiazine and showed that serum silver concentrations reached 50 μg/l within six hours, with maximal concentrations of 310 μg/l. Urinary silver excretion after one day ranged from 100 to 400 μg/day (normal range: 1 μg). Silver absorbed from wound sites enters the systemic circulation and is distributed to the liver, kidney, brain, eye, and other organs. A biopsy from a patient who had been treated with SSD for eight days revealed 14 μg Ag/g liver weight. The kidney plays a major role in eliminating silver from the human body, but it is unclear whether it is susceptible to toxicity. There was no correlation between serum silver or renal function in patients with burn injuries who received 180 μg/kg SSD and had serum concentrations over 100 μg/l silver. Patients treated with SSD had silver granules in the renal tubule, but most were bound in lysosomes or as plasma–protein complexes.

10.12 Clinical Efficacy of the Silver-Containing Dressings

Experimental studies have demonstrated that silver induces metallothioneins (MT-1 and MT-2) in epithelial cells at the wound margin and in dermal fibroblasts [25]. MTs play an essential part in the intracellular uptake and metabolism of zinc and copper in actively proliferating cells. They bind to and are induced by several xenobiotic metals, including silver, cadmium, gold, and mercury. MTs are cysteine-rich low-molecular-weight proteins with a possible mitogenic action. They also protect tissues against the toxic effects of metals such as cadmium and mercury [2, 25].

Accumulation of silver in the rat wound was associated with increased zinc, presumably reflecting higher levels of zinc metalloenzymes. This was, in turn, associated with increased epidermal cell proliferation and improved re-epithelialization.

Jørgensen et al. [18] compared a sustained-release silver foam dressing with a foam dressing without silver in patients with venous or mixed venous/arterial leg ulcers and signs of stalled or delayed healing caused by suspected critical colonization. The silver foam dressing decreased wound size by 50%, with the comparators reducing wound size by 30%. There was also a statistically significant increase in the presence of normal granulation tissue in 68% of the sustained-release silver foam-treated patients and 50% of the local best practice group. The study also revealed that the sustained-release silver foam dressing showed statistically significant advantages in odor reduction and pain relief.

Finally, it should be pointed out that other than occasional references to argyria-like discolorations in wounds treated with high silver-releasing dressings, there is no evidence that the excess silver exerts any undesirable effects within the wound bed or systemically [23].

10.13 Summary

The antibacterial efficacy of silver is well documented in laboratory tests. When combined with alginate wound dressings, silver can help control bacteria growth on the wound site, and to a certain extent, accelerate wound healing. Silver in various forms has been added to wound dressings to act as an effective antimicrobial agent. Because of the diversity in both the types of silver compounds and choices of base materials, large varieties of silver-containing wound dressings are now commercially available. These products differ in their overall silver contents, the contact areas for silver, the rate and amount of silver released, and the absorption capacities of the dressing. They can be used to treat wounds with different levels of infection and with different amounts of exudate, providing the wound management practice with an effective means of patient care. Among the many types of silver-containing wound dressings, silver-containing alginate wound dressings have become one of the most effective products in the management of wound infection, through the combination of the antimicrobial efficacy of silver and the gelling ability of alginate fibers.

References

1 Bleehan, S.S., Gould, D.J., and Harrington, C.I. (1981). Occupational argyria: light and electron microscopic studies and x-ray microanalysis. *Br. J. Dermatol.* 104: 19–26.

2 Bremner, I. and Beattie, J.H. (1990). Metallothionein and the trace metals. *Annu. Rev. Nutr.* 205: 25–35.

3 Buckley, W.R., Oster, C.F., and Fassett, D.W. (1965). Localized argyria II. Chemical nature of the silver containing particles. *Arch. Dermatol.* 92: 697–704.

4 Charley, R.C. and Bull, A.T. (1979). Bioaccumulation of silver by a multispecies population of bacteria. *Arch. Microbiol.* 123: 239–244.

5 Clarke, A.J. (1937). General pharmacology. In: *Handbuch Der Experimentelle Pharmakologie* (ed. A. Heffter), 46–63. Berlin: Springer.

6 Coombs, C.J., Wan, A.T., Masterton, J.P. et al. (1992). Do burns patients have a silver lining? *Burns* 18 (3): 179–184.

7 Deitch, E.A., Marino, A.A., Malakanok, V. et al. (1987). Silver nylon cloth: *in vitro* and *in vivo* evaluation of antimicrobial activity. *J. Trauma* 27 (3): 301–304.

8 Demling, R.H. and DiSanti, L. (2001). Effects of silver on wound management. *Wounds* 13 (Suppl. A): 5–15.

9 Di Vincenzo, G.D., Giordiano, C.J., and Schreiver, L.S. (1985). Biologic monitoring of workers exposed to silver. *Int. Arch. Occup. Environ. Health* 56 (3): 207–215.

10 Dolmer, M., Larsen, K., and Jensen, K. (2004). In vitro silver release profiles for various antimicrobial dressings. In: *Paper Presented at the 2nd World Union of Wound Healing Societies Meeting, Paris.*

11 Ersek, R.A. and Denton, D.R. (1988). Cross-linked silver-impregnated skin for burn wound management. *J. Burn Care Rehabil.* 9 (5): 476–481.

12 Feng, Q.L., Wu, J., Chen, G.Q. et al. (2000). A mechanistic study of the antibacterial effect of silver ions on Escherichia coli and Staphylococcus aureus. *J. Biomed. Mater. Res.* 52: 662–668.

13 Furr, J.R., Russell, A.D., Turner, T.D. et al. (1994). Antibacterial activity of Actisorb plus, Actisorb and silver nitrate. *J. Hosp. Infect.* 27 (3): 201–208.

14 Hamilton-Miller, J.M., Shah, S., and Smith, C. (1993). Silver sulphadiazine: a comprehensive in vitro re-assessment. *Chemotherapy* 39: 405–409.

15 Hoffmann, S. (1984). Silver sulfadiazine: an antibacterial agent for topical use in burns. A review of the literature. *Scand. J. Plast. Reconstr. Surg.* 18: 119–126.

16 Hollinger, M.A. (1996). Toxicological aspects of silver pharmaceuticals. *Crit. Rev. Toxicol.* 26: 255–260.

17 Hostynek, J.J., Hinz, R.S., Lorence, C. et al. (1993). Metals and the skin. *Crit. Rev. Toxicol.* 23 (2): 171–235.

18 Jørgensen, B., Price, B., Andersen, K.E. et al. (2005). The silver-releasing foam dressing, Contreet foam, promotes faster healing of critically colonized venous leg ulcers: a randomized, controlled trial. *Int. Wound J.* 2 (1): 64–73.

19 Kirsner, R.S., Orsted, H., and Wright, J.B. (2001). Matrix metalloproteinases in normal and impaired wound healing: a potential role of nanocrystalline silver. *Wounds* 13 (2): 4–12.

20 Lansdown, A.B.G. (1995). Physiological and toxicological changes in the skin resulting from the action and interaction of metal ions. *CRC Crit. Rev. Toxicol.* 25: 397–462.

21 Lansdown, A.B.G. (2002). Silver 1: its antimicrobial properties and mechanism of action. *J. Wound Care* 11: 125–131.

22 Lansdown, A.B.G. (2004). A review of silver in wound care: facts and fallacies. *Br. J. Nurs.* 13 (Suppl): 6–19.

23 Lansdown, A.B. and Williams, A. (2004). How safe is silver in wound care? *J. Wound Care* 13 (4): 131–136.

24 Lansdown, A.B.G., Sampson, B., Laupattarakasem, P. et al. (1997). Silver aids healing in the sterile wound: experimental studies in the laboratory rat. *Br. J. Dermatol.* 137: 728–735.

25 Lansdown, A.B.G., Sampson, B., and Rowe, A. (1999). Sequential changes in trace metal, metallothionein and calmodulin concentrations in healing skin wounds. *J. Anat.* 195: 375–386.

26 Lansdown, A.B.G., Williams, A., Chandler, S. et al. (2005). Silver absorption and antibacterial efficacy of silver dressings. *J. Wound Care* 14 (4): 155–160.

27 Marshall, W. and Schneider, R.P. (1977). Systemic argyria secondary to topical silver nitrate. *Arch. Dermatol.* 113: 1077–1079.

28 Meaume, S., Senet, P., and Dumas, R. (2002). Urgotul: a novel non-adherent lipido-colloid dressing. *Br. J. Nurs.* 11 (16): 42–50.

29 Modak, S.M. and Fox, C.L. (1973). Binding of silver sulfadiazine to the cellular components of Pseudomonas aeruginosa. *Biochem. Pharmacol.* 22 (19): 2391–2404.

30 Ovington, L.G. (2001). Nanocrystalline silver: where the old and familiar meets a new frontier. *Wounds* 13 (Suppl. B): 5–10.

31 Qin, Y. (2016). *Medical Textile Materials*. Cambridge: Woodhead Publishing.
32 Qin, Y. (2022). *Silver Containing Functional Wound Dressings*. Beijing: China Textile Press.
33 Russell, A.D. and Hugo, W.B. (1994). Antimicrobial activity and action of silver. In: *Progress in Medicinal Chemistry* (ed. G.P. Ellis and D.K. Luscombe), 112–135. New York: Elsevier Science.
34 Thomas, S. and McCubbin, P. (2003). A comparison of the antimicrobial effects of four silver-containing dressings on three organisms. *J. Wound Care* 12 (3): 101–107.
35 Thomas, S. and McCubbin, P. (2003). An in vitro analysis of the antimicrobial properties of 10 silver-containing dressings. *J. Wound Care* 12 (8): 305–308.
36 Tredget, E.E., Shankowsky, H.A., Groeneveld, A. et al. (1998). A matched-pair, randomized study evaluating the efficacy and safety of Acticoat silver-coated dressing for the treatment of burn wounds. *J. Burn Care Rehabil.* 19 (6): 531–537.
37 Wan, A.T., Conyers, R.A., Coombs, C.J. et al. (1991). Determination of silver in blood, urine and tissues of volunteers and burn patients. *Clin. Chem.* 37: 1683–1687.
38 Wells, T.N., Scully, P., Paravicini, G. et al. (1995). Mechanisms of irreversible inactivation of phosphomannose isomerases by silver ions and flamazine. *Biochemistry* 34: 7896–7903.
39 Wright, J.B., Lam, K., and Burrell, R.E. (1998). Wound management in an era of increasing bacterial antibiotic resistance: a role for topical silver treatment. *Am. J. Infect. Control* 26 (6): 572–577.
40 Wright, J.B., Lam, K., Hansen, D. et al. (1999). Efficacy of topical silver against fungal burn wound pathogens. *Am. J. Infect. Control* 27 (4): 344–350.
41 Wright, J.B., Lam, K., Buret, A.G. et al. (2002). Early healing events in a porcine model of contaminated wounds: effects of nanocrystalline silver on matrix metalloproteinases, cell apoptosis and healing. *Wound Repair Regen.* 10 (3): 141–151.
42 Yin, H.Q., Langford, R., and Burrell, R.E. (1999). Comparative evaluation of the antimicrobial activity of Acticoat antimicrobial barrier dressing. *J. Burn Care Rehabil.* 20 (3): 195–200.

A

Appendix A: List of Silver Containing Wound Dressings

Product name	Manufacturer	Base material	Silver compound
		Alginate	
Acticoat Absorbent	Smith & Nephew	Alginate fiber with PE film	Nanocrystalline silver
Algicell Ag	Derma Sciences	Calcium alginate fiber	1.4% silver
Algidex Ag	DeRoyal	Calcium alginate fiber with foam backing	Ionic silver
Algisite Ag	Smith & Nephew	Calcium alginate fiber	Silver impregnated
Askina Calgitrol Ag	B. Braun	Calcium alginate fiber with foam backing	Ionic silver
Askina Calgitrol THIN	B. Braun	Thin alginate sheet	Ionic silver
Askina Calgitrol Paste	B. Braun	Alginate in paste form	Ionic silver
Invacare Silver Alginate	Invacare	Alginate/CMC fiber	Silver sodium hydrogen zirconium phosphate
Maxorb Extra Ag	Medline	Alginate/CMC fiber	Silver sodium hydrogen zirconium phosphate
Melgisorb Ag	Mölynlycke	Alginate/CMC fiber	Silver sodium hydrogen zirconium phosphate
Restore Calcium Alginate	Hollister Woundcare	Calcium alginate	Ionic silver
SeaSorb Ag	Coloplast	Alginate/CMC fiber	Silver sodium hydrogen zirconium phosphate

(*continued*)

Alginate Fibers and Wound Dressings: Seaweed Derived Natural Therapy, First Edition. Yimin Qin.

Product name	Manufacturer	Base material	Silver compound
Silvercel; Silvercel Non-Adherent	Systagenix	Alginate/CMC fiber with non-adherent contact layer	Elemental silver-coated nylon fibers
Silverlon Calcium Alginate	Argentum Medical	Calcium alginate fiber	Metallic silver plated nylon mesh core
Sorbsan Silver Flat; Sorbsan Silver Packing; Sorbsan Silver Plus NA	Aspen Medical	Calcium alginate fiber plus viscose pad or film backing	1.5% ionic silver
Suprasorb A + Ag	Activa Healthcare	Calcium alginate fiber	Silver unspecified
Tegaderm Alginate Ag	3 M	Alginate/CMC fiber	Silver sodium hydrogen zirconium phosphate
UrgoSorb Silver	Urgo	Alginate/CMC fiber	Silver sodium hydrogen zirconium phosphate
		Collagen	
BIOSTEP Ag	Smith & Nephew	Collagen and ethylenediaminetetracetic acid	Silver chloride
COLACTIVE collagen with silver	Smith & Nephew	Collagen and alginate	Silver lactate
Covaclear Ag Hydrogel	Covalon	Collagen-based hydrogel	Silver (form not specified)
Promogran Prisma	Systagenix	Collagen and oxidized regenerated cellulose	1% silver (silver-ORC compound)
Puracol Plus Ag^+	Medline	Collagen	Silver chloride
		Cream	
Flamazine	Smith & Nephew	Cream base	SSD
		Fabric	
ACTICOAT; ACTICOAT 7	Smith & Nephew	Rayon and polyester core	Nanocrystalline silver
Actisorb Silver 220	Systagenix	Activated charcoal cloth in nylon fabric sleeve	Impregnated with silver
Atrauman Ag	Paul Hartmann	Polyester wound contact layer	Impregnated with silver

(*continued*)

Product name	Manufacturer	Base material	Silver compound
Physiotulle Ag	Coloplast	Knitted polyester net with hydrocolloid particles, petrolatum	SSD
Restore Contact Layer Dressing with Silver	Hollister Woundcare	Non-adherent dressing	Silver sulfate
Silverlon Wound Contact Dressings	Argentum	Nylon fabric	Silver coated
Silverseal Contact Dressing	Derma Sciences	Knitted fabric	99.1% elemental silver and 0.9% silver oxide
Tegaderm Ag Mesh	3 M	Gauze	Silver sulfate
Urgotul Duo Silver	Urgo	Polyester mesh with lipido-colloid coating; viscose backing	Impregnated with silver salt
Urgotul SSD	Urgo	Polyester mesh with lipido-colloid coating	Impregnated with SSD
Vliwaktiv Ag	Lohmann and Rauscher	Activated charcoal dressing	Impregnated with silver (form not specified)
		Film	
Arglaes Film Island; Arglaes Island	Medline	Film dressing; Island has an alginate pad	Ionic silver
		Foam	
ACTICOAT Moisture Control	Smith & Nephew	Polyurethane wound contact layer, foam core, and film backing	Nanocrystalline silver coated
ALLEVYN Ag Adhesive; ALLEVYN Ag Heel	Smith & Nephew	Adhesive foam, film backing	SSD
ALLEVYN Ag Non-Adhesive	Smith & Nephew	Non-adhesive foam, shaped for heel	SSD
Avance	Mölnlycke	Non-adhesive foam dressing	Impregnated with silver
Avance A	Mölnlycke	Adhesive foam dressing	Impregnated with silver
Biatain Ag	Coloplast	Adhesive foam, film backing	Impregnated with silver
Mepilex Ag	Mölnlycke	Soft silicone contact layer, foam core, film backing	Silver (form not specified)

(continued)

Product name	Manufacturer	Base material	Silver compound
Optifoam	Medline	Foam pad	Silver (form not specified)
Polymem Silver	Ferris Manufacturing Corp	Foam dressing, starch, and glycerin	Impregnated with silver
Urgocell Silver	Urgo	Foam core with lipido-colloid contact layer and film backing	Silver impregnated
		Gauze	
Tegaderm Ag	3 M	Non-woven mesh/gauze	Impregnated with silver sulfate
Urgotul SSD	Urgo Medical	Polyester mesh with hydrocolloid and petroleum jelly	Impregnated SSD
		Hydrocolloids	
Contreet Hydrocolloid	Coloplast	Impregnated hydrocolloid with vapor-permeable backing	Silver (form not specified)
Silverseal Hydrocolloid	Alliqua	Hydrocolloid dressing	Silver (form not specified)
Sureskin Silver	EuroMed	Hydrocolloid dressing	Sodium hydrogen zirconium phosphate
		Hydrofiber	
AQUACEL Ag	ConvaTec	Hydrofiber	1.2% ionic silver
		Hydrogel	
AquaMed Hydrogel Sheet with Silver	AquaMed Technologies	Hydrogel	Elemental silver
Gentell Hydrogel Ag	Concept Health	Hydrogel	SSD
Silvasorb Gel	Medline	Hydrogel	Silver (form not specified)
Silverseal Hydrogel	Alliqua	Hydrogel with fibers	Silver coated
		Powder	
Arglaes Powder	Medline	Alginate powder	Ionic silver (form not specified)

B

Appendix B: Answers to Commonly Asked Questions About Alginate Wound Dressings

B.1 What Are Alginic Acid, Sodium Alginate, and Calcium Alginate?

Alginic acid is a polymeric acid composed of two types of monomers, i.e. mannuronic acid (M) and guluronic acid (G). The carboxylic acid groups along the polymeric chain can combine with metal ions to form alginate salts, such as sodium alginate and calcium alginate. In brown seaweeds where alginate is the main structural component of cell walls, alginate exists mainly as calcium, magnesium, and sodium salts. During the extraction process, they are all converted into the water-soluble sodium alginate and are separated from the seaweed biomass following dissolution, precipitation, purification, and drying operations. Figure B.1 shows the chemical structures of mannuronic acid (M) and guluronic acid (G).

B.2 What Do M and G Mean With Alginate Fibers and Wound Dressings?

The two monomers in alginic acid are mannuronic acid and guluronic acid, which are isomers with different gelling properties. These two monomers are abbreviated as M and G, respectively. Naturally occurring alginic acid in brown seaweeds has a

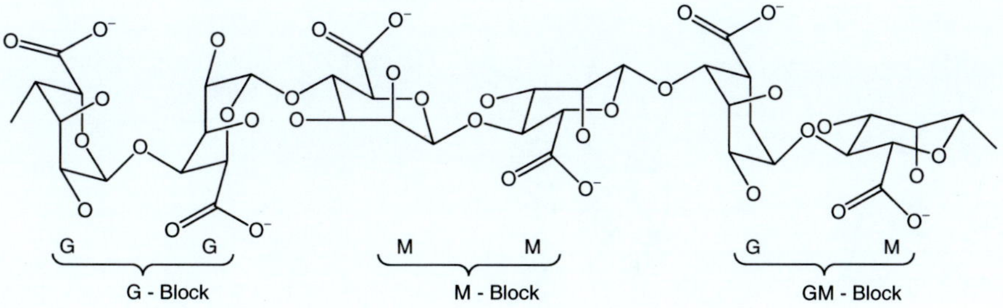

Figure B.1 Chemical structures of mannuronic acid (M) and guluronic acid (G).

Alginate Fibers and Wound Dressings: Seaweed Derived Natural Therapy, First Edition. Yimin Qin.

very high molecular weight in the range of 20 000–200 000, corresponding to a degree of polymerization of between 100 and 1000. These monomers are arranged in blocks such as GG, MM, and GM, where the GG blocks are responsible for stronger binding with divalent metal ions such as calcium ions, while the MM blocks have lower binding capabilities. High G alginates are those with relatively high content of guluronic acid and are able to form strong gels when in contact with calcium ions, while on the other hand, high M alginates usually form soft hydrogels.

B.3 What Are the Differences Between Alginate Wound Dressing and Calcium Alginate Wound Dressing?

Alginate wound dressings are also commonly known as calcium alginate wound dressings. As a polymeric acid, alginic acid can combine with metal ions to form alginate, which is a polymeric salt. During the wet spinning process, water-soluble sodium alginate is dissolved in water to form the spinning solution, which is extruded through spinneret into a coagulation bath commonly made of aqueous calcium chloride solution, although other aqueous solutions of divalent metal ions such as zinc chloride and copper chloride can also be used to precipitate alginate to form fibers. In general, divalent metal ions, such as calcium, zinc, and copper, can form cross-linked gel structure, allowing the formation of fibers during wet spinning. Calcium chloride is used mainly because calcium ions are safe for the human body, which is suitable for the resultant calcium alginate fibers to be used in medical and hygiene applications. Figure B.2 shows the formation of calcium alginate hydrogel upon the exchange between sodium ions and calcium ions.

Because of the incomplete ion exchange during coagulation, the fibers obtained from the wet-spinning process often contain some sodium ions. Sometimes, it is deliberate to make fibers that contain a significant percentage of sodium alginate in order to make the fibers more absorbent. Zinc and copper ions are also added into the fibers to obtain specific functional properties such as antimicrobial efficacy. These various types of fibers are generally known as alginate fibers, although in the medical and hygiene industries, calcium alginate fibers are mainly used.

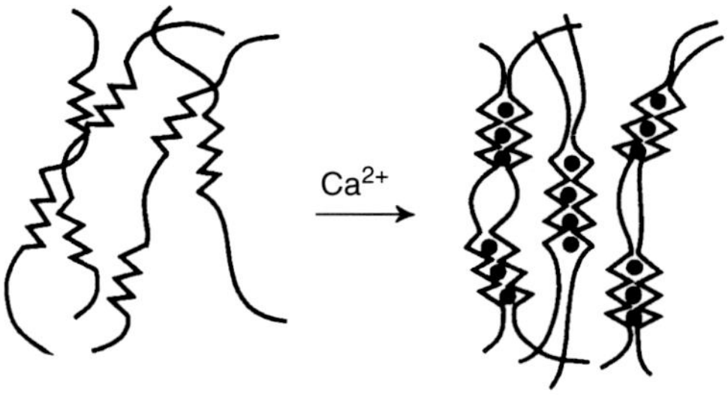

Figure B.2 An illustration of the formation of calcium alginate gel.

B.4 How Can Calcium Alginate Dressing Form Hydrogel on Contact With Wound Exudate?

When calcium alginate fibers in the nonwoven dressing are in contact with wound exudate, the sodium ions in the exudate exchange with calcium ions in the fibers, resulting in the formation of fibers that contain both calcium and sodium alginate. Since sodium alginate is water soluble, when more and more calcium ions are replaced by sodium ions, the fibers pick up more and more water into the fiber structure, which is accompanied by the swelling of the fibers to form a fibrous hydrogel. During application on exuding wounds, these dressings are capable of absorbing a large amount of wound exudate and forming a moist healing environment due to the conversion of calcium alginate fibers to calcium sodium alginate fibers.

B.5 What Role Does Alginate Wound Dressings Play in "Moist Healing"?

The birth and widespread acceptance of the "moist healing" mechanism in wound management helped to promote the applications of alginate wound dressings. In 1962, Dr. George Winter of the University of London published his findings in Nature concerning the formation of scab and the rate of epithelialization of superficial wounds in the skin of the young domestic pig. It was found that wounds that are kept in a moist condition heal faster than being kept in a dry condition. Under the guidance of this theory, modern wound dressings such as hydrocolloids, polyurethanes, and alginate wound dressing are designed to main a moist but not wet condition on the wounds, resulting in a fundamental change in the research, development, and applications of modern functional wound dressings. Since the 1980s, many types of alginate wound dressings have been developed and launched in the world market. Due to their excellent properties, alginate wound dressings gained widespread acceptance as an excellent product for the management of wounds with medium to high levels of exudate.

B.6 What Are the Main Applications of Alginate Wound Dressings?

Alginate wound dressings are by their chemical and physical nature highly absorbent when in contact with wound exudate, making them suitable for managing exuding wounds. The fibers can be processed into different forms in order to meet clinical requirements concerning different types of wounds. The main type of alginate wound dressing is made of nonwoven fabrics which are produced

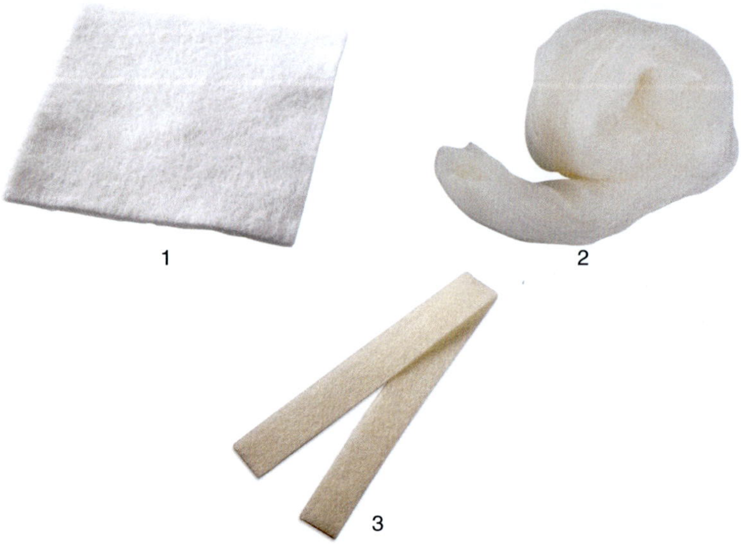

Figure B.3 Main varieties of alginate wound dressings, 1. nonwoven fabric; 2. rope; 3. strip.

by first opening the fibers and card them into a thin sheet of loose fibers which are then cross-lapped into a thicker web before being needled to form nonwoven alginate fabric. This type of flat sheet structure as shown in Figure B.3 (1) is suitable for covering flat open wounds. Alternatively, the fibers can be carded and then combined into a rope with a porous loose structure, which is used to fill cavity wounds. The nonwoven fabric can also be cut into narrow strips which are also used to fill deep cavities. Table B.1 shows the common sizes for alginate dressings.

Nonwoven alginate fabric can be combined with other types of wound management materials to make plasters that are easier to apply. Figure B.4 shows an alginate plaster with a piece of nonwoven alginate as the core and hydroentangled viscose nonwoven fabric as the outer layer, which is coated with pressure-sensitive adhesive. The outer layer can also be made of polyurethane film, membrane, and foam to offer more functional properties, such as regulating water evaporation, and preventing bacteria invasion. These novel properties are suited for the management of many types of wounds, as illustrated in Figure B.5.

Table B.1 Common sizes for alginate dressings.

Product	Size
Flat dressing	5 cm × 5 cm, 5 cm × 10 cm, 7.5 cm × 12.5 cm, 10 cm × 10 cm, 10 cm × 20 cm, 20 cm × 20 cm, 20 cm × 30 cm, 30 cm × 30 cm
Cavity filler	1 g/15 cm, 1 g/30 cm, 2 g/30 cm

Figure B.4 Alginate plaster.

Figure B.5 Clinical applications of alginate wound dressings.

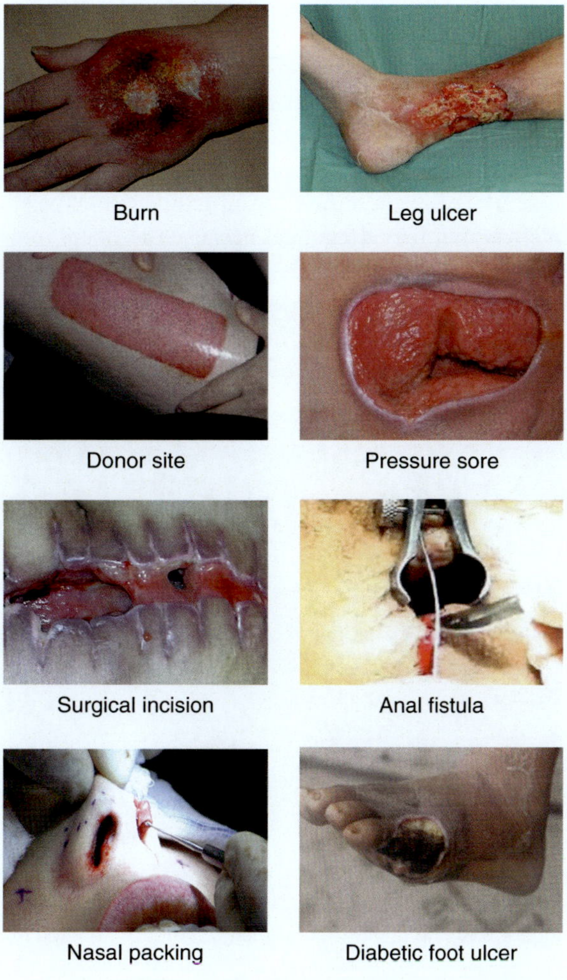

B.7 What Is the Absorption Mechanism of Alginate Wound Dressing?

Commercial alginate wound dressings are mostly made through nonwoven processes where the fibers are arranged in loose forms. The nonwoven structure is highly porous and there are a lot of capillary spaces between fibers that can hold water when the dressing is in contact with wound exudate. Once the exudate enters into the dressing, the fibers swell following the exchange between calcium ions in the fiber and sodium ions in the exudate. This swelling causes the closure of the porous structure, thereby blocking the lateral spreading of liquid, resulting in the "gel blocking" mechanism of alginate wound dressings which is known to prevent the maceration of skin on the edge of the wounded area. The vertical rise of wound exudate is not affected because during the production process, the needling of the fabric creates vertical channels that can facilitate the diffusion of wound exudate. Figure B.6 offers an illustration of the structure of needled nonwoven fabric.

B.8 What Is the Reason That Calcium Alginate Fibers Do Not Gel in Pure Water?

For alginate fibers, gelling means the expansion of fiber diameter following the diffusion of water into the fiber structure. When calcium alginate fibers are placed in plain water, they do not gel because the fibers remain in a closely intact structure as the polymeric molecules are bound by hydrogen bonds through the –OH groups along the polymer chain. When the contact solution contains sodium ions, as is the case for wound exudate, the sodium ions would exchange with calcium ions in the calcium alginate fibers, and as a result, part of the fibers are converted into sodium

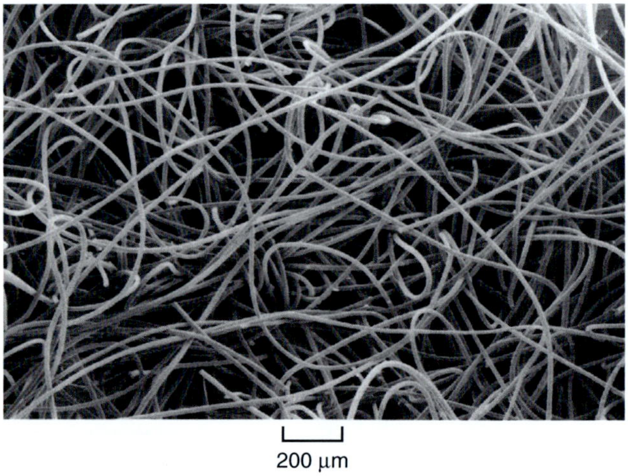

Figure B.6 An illustration of the structure of needled nonwoven fabric.

alginate. Since sodium alginate is water soluble, a large amount of water is drawn into the fibers, turning them into fibrous hydrogel which can provide a moist healing environment when applied on exuding wounds.

B.9 What Are the Differences Between the Absorption of Wound Exudate by Cotton Gauze and Alginate Wound Dressings?

Figure B.7 compares the absorption of liquid by alginate wound dressing and cotton gauze. In the cotton gauze, the liquid is mainly held in the spaces between yarns by physical force, which can easily spread along the gauze, and the water is also easy to evaporate. On the other hand, when alginate fibers absorb water, the water is absorbed into the fibers where it is combined with the carboxylic acid groups to form a hydrogel. Since the water is chemically combined with alginate, it is not easy to evaporate and a moist healing environment can be facilitated on the wound surface.

B.10 What Are the Differences Between High G and High M Alginate Fibers When They Are Applied on Exuding Wounds?

As a polymeric acid, alginate has ion-exchange properties and in the case of wound dressings, the ion-exchange is mainly concerned with calcium and sodium ions. The exchange of calcium alginate fiber with sodium ions in the contacting solution is guided by the following equation:

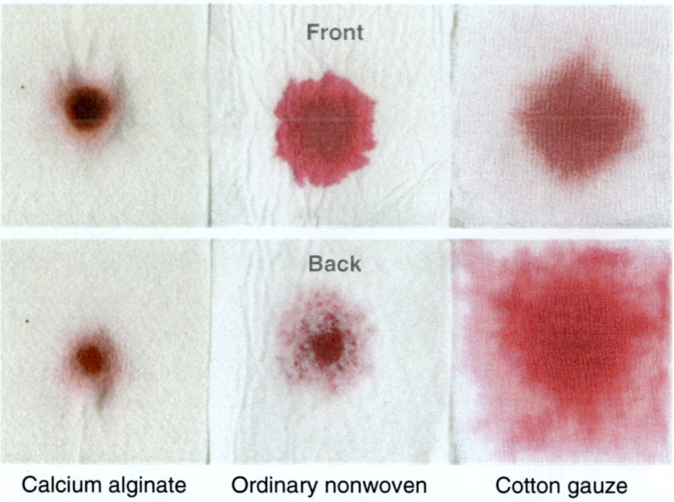

Figure B.7 An illustration of the absorption of liquid by calcium alginate dressing, ordinary nonwoven fabric, and cotton gauze.

Table B.2 The ion-exchange coefficient for two types of alginate.

Metal ions	M/G ratio	
	High M, M/G = 1.60	High G, M/G = 0.45
$Ca^{2+} - Na^{+}$	7.5	20

$$K = \frac{[\text{Calcium ions in the fiber}][\text{Sodium ions in the solution}]^2}{[\text{Sodium ions in the fiber}]^2[\text{Calcium ions in the solution}]}$$

The ion-exchange coefficient K is dependent on the guluronic acid and mannuronic acid content of alginate, with high G alginate having relatively high exchange coefficient. As can be seen in Table B.2, the G and M contents of alginate have a significant effect on the exchange coefficient of alginate. When high M alginate fibers are applied on exuding wounds, the sodium ions in the exudate can easily replace the calcium ions in the fiber, resulting in the rapid formation of a hydrogel on the wound surface. On the other hand, high G alginate fibers are more difficult to form hydrogel, mainly because of the low ion exchange coefficient between calcium ions in the fibers and sodium ions in the wound exudate.

B.11 Can Sodium Alginate Be Absorbed by the Body When Calcium Alginate Fibers Are Converted Into Sodium Alginate Upon Contact With Wound Exudate?

Alginate is a marine biopolymer that cannot be digested by the human body since there is no enzyme in the human system to digest alginate. In the food industry, alginate is known as a dietary fiber that can benefit the digestive system mainly because it cannot be digested by the digestive system. When applied on the wound surface, the calcium alginate fibers are converted into sodium alginate through ion exchange and as the wound heals, the alginate fibers are detached from the wound surface as part of the healing process.

B.12 Alginate Wound Dressings Are Divided Into Wet Integral and Wet Dispersible, What Does These Two Types Mean?

The wet integrity of alginate wound dressings are mainly caused by the variations in the chemical structure of alginic acid. Alginate with a high degree of guluronic acid is capable of strong ionic binding and the resultant fibers are stable in solution A and are integral when applied on exuding wounds. On the other hand, alginate fibers made from alginate with a high content of mannuronic acid are capable of a high degree of swelling, and as a result, wound dressings made from high M alginate fibers lose the nonwoven structure easily when in contact with wound exudate. In the test method of British Pharmacopoeia, wet integrity means that the dressing can retain its nonwoven structure when shaken in 40 times its own weight of solution A, while wet dispersible means that the dressing loses its nonwoven structure under the same circumstances.

B.13 How Can Alginate Wound Dressings Reduce Pain?

Pain is caused by mechanical stimulation such as knife cutting, stick beating, physical factors such as electrical current, high temperature, and chemical factors such as strong acid and strong bases. Pain is common in clinical practices. Alginate wound dressings can relieve pain through its in-situ formation of hydrogel on contact with exuding wounds to provide a moist hydrogel on the wound which can reduce friction and offer a protective layer to the nerve endings exposed to the wound. These dressings can be easily removed, which offers an additional benefit to pain relief.

B.14 Are There Any Inappropriate Consequences for the Residue Alginate Fibers Left on the Wound Surface?

As a loosely constructed nonwoven fabric, alginate wound dressings have loose fibers on the surface of the dressing, which would shed on the wound surface when applied on exuding wounds. In addition, when the dressing is peeled off during removal, some of the fibers in the dressing would be left behind. However, these residue fibers are harmless since first, they are nontoxic to the human body, and second, as the wound heals, the dead tissue on the wound would be detached from the body together with the fibers as a part of the metabolic process for skin regeneration. Many clinical applications have proven that the loose fibers left on wound are harmless to the human body.

B.15 Are There Any Differences Between the Alginate Wound Dressings Under Different Brands?

Yes, the alginate wound dressings under different brands differ considerably in their chemical structures and physical properties, mainly due to the fact that alginates are different in their monomer composition in that the contents of α-L-guluronic acid (M) and β-D-mannuronic acid (M) are considerably different depending on the types of brown seaweeds where they are extracted from. In addition, the fiber composition and nonwoven construction are also different for different types of alginate wound dressings. Table B.3 summarizes the product characteristics of the main types of alginate wound dressings in the international market.

Table B.3 Product characteristics of the main types of alginate wound dressings in the international market.

Brand name	Distributor	Product characteristics
Algisite	Smith & Nephew	Needle-punched high M calcium alginate
Algosteril	Beiersdorf	Needle punched high G calcium alginate
Comfeel SeaSorb	Coloplast	Freeze dried alginate foam
Kaltogel	Convatec	Needle-punched high M calcium alginate
Kaltostat	Convatec	Needle punched high G calcium/sodium alginate
Melgisorb	Molnlycke	Needle-punched high M calcium alginate
Sorbsan	Maersk	Nip rolled high M calcium alginate
Tegagel	3 M Health Care	Hydro-entangled high M calcium alginate

B.16 In the Manufacturing Processes, What Do Nip Rolling, Needle Punching, and Freeze Drying Mean?

During the manufacture of alginate wound dressings, the alginate fibers are carded on a nonwoven processing machine to form a thin web of fibers which is then cross-lapped to form a thick layer of fibers. At this stage, the layer of fibers is either pressed to form a loosely connected fabric that is further cut into small pieces of dressings or it is punched by needles to form an integral nonwoven fabric before being processed into dressings. In the case of loosely formed fabric, the resultant wound dressing can be easily washed off the wound at the end of application, while in the case of integral needle-punched nonwoven dressing, it can be lifted off the wound as a while piece.

Freeze-dried alginate wound dressing is made by dissolving sodium alginate to form a viscous liquid and placing the solution in a flat-bottomed container before calcium ions are added to form calcium alginate gel. This gel is then freeze-dried to form a porous foam material, which is then cut into pieces and sterilized to produce wound dressings with high absorption capacity.

B.17 Can Alginate Wound Dressings Be Used in Combination With Topical Medicines Such as Anti-inflammatory Drugs Like Iodine?

Alginate wound dressings can be used in combination with iodine and any other anti-inflammatory drugs to protect the wound against infection.

B.18 Can Alginate Wound Dressings Be Used on Infected Wounds?

Alginate wound dressings can be used on infected wounds, especially when they contain antimicrobial agents such as silver ions.

B.19 Can Alginate Wound Dressings Be Cut Into Pieces Before Being Applied to Wounds?

In order to fit into the wound, alginate wound dressings can either be folded or cut into small pieces that can be applied to wound of various depths and sizes.

B.20 What Should Be Done When Alginate Wound Dressings Adhere to the Wound Surface?

When alginate dressing adheres to the wound bed, the easy way to remove it is to apply warm saline on the dressing. Once the fibers swell due to the ion exchange between calcium ions in the dressing and sodium ions in saline, they become soft and can be easily rinsed off.

B.21 Some Patients Experience Granulation Edema When Applied With Alginate Wound Dressings, Is This Related to the Release of Calcium Ions by the Dressing?

Wound healing is a complicated process. Some individual patients may experience granulation edema because the ionic content in the wound bed is different from the rest of the tissue. The dressings should be frequently changed in order to maintain a fresh environment at the wound surface. For the calcium ions released during the ion exchange process, the majority should remain in the dressing structure. Some may diffuse into the tissue fluid.

B.22 What Are the Clinical Efficacy of Alginate Wound Dressings When Used for Pressure Sore Wounds?

Many literature studies have shown that when alginate wound dressings are used on pressure sore wounds, they can reduce the number of dressing changes and related nursing time and direct cost, hence improve overall nursing efficiency. Compared to traditional dressings, the clinical benefits include reduction of pain during dressing change, improvement in healing rate, and reduction in overall nursing time. The alginate wound dressings are safe, effective, and comfortable to use, they can promote the healing of phases III–IV pressure sore wounds, especially when they are used in combination with transparent semi-permeable film dressings, when they can effectively promote wound tissue granulation and capillary formation. In general, when used on pressure-sore wounds, the alginate wound dressings cause minimal wound tissue damage and can improve the quality of wound healing.

B.23 What Are the Clinical Efficacy of Alginate Wound Dressings When Used for Leg Ulcer Wounds?

When alginate wound dressings are used on leg ulcer wounds, the clinical benefits include improved reduction in wound surface, pain level, and bleeding during dressing change as compared to traditional therapy. The dressing does not adhere to the wound bed and it can be easily removed without causing pain to patients.

B.24 What Are the Clinical Efficacy of Alginate Wound Dressings When Used for Diabetic Foot Ulcer Wounds?

For diabetic foot ulcer wounds, the application of alginate wound dressings can help to facilitate a moist healing environment that benefits the healing process. The healing time can be reduced on average by 5–7 days. In addition, alginate wound dressings can accelerate the migration of immune cells and promote phagocytes to digest dead tissue, bacteria, and other microorganisms. The high absorption capacity of alginate wound dressings is useful in containing the dead tissue on the wound surface and also donating moist to the wound to soften hard eschar.

B.25 What Are the Clinical Efficacy of Alginate Wound Dressings When Used for Burn Wounds?

Alginate wound dressings can be used on burn wounds with many clinical benefits. In general, alginate wound dressings can absorb a large amount of wound exudate while at the same time forming a soft hydrogel on wound surface in situ. For burn wounds, frequent dressing change is not desirable since it not only increases pain, it also introduces exposure to wound infection. Alginate wound dressings can provide a moist healing environment that reduces pain, minimizes infection, and also promotes healing.

B.26 What Are the Clinical Efficacy of Alginate Wound Dressings When Used for Anal Fistula Wounds?

When alginate wound dressings are used for anal fistula wounds, they are mainly used as a cavity filler, which can reduce pain, reduce the level of bleeding, and improve the absorption of wound exudate. In the area surrounding anal fistula wounds, the gelling properties of alginate wound dressings are particularly useful in that the nonwoven structure can immobilize the exudate and the microorganisms within the liquid. The hydrophilic nature of the dressing can reduce water evaporation and help to maintain a moist environment that is highly beneficial to the patient.

B.27 Does Alginate Wound Dressing Possess Hemostatic Properties?

Calcium alginate dressings have excellent hemostatic properties. Many clinical studies have confirmed that bleeding stopped within five minutes of applying the dressing. It has also been shown that the excellent hemostatic properties of alginate dressings are due mainly to two reasons, i.e. blood coagulation and enhancement of platelet activity. Because of the high absorption capacity and the ability to release calcium ions, the hemostatic properties of alginate dressings are better than most surgical dressings. In addition, zinc alginate dressing possesses better hemostatic properties than calcium alginate dressing.

B.28 Where Are the Seaweeds Used in the Production of Alginate Wound Dressings Come From?

Marine seaweeds are divided into red, green, and brown seaweeds. Alginate is extracted from brown seaweeds which mainly grow in cold environments such as the north Atlantic Ocean and southern Pacific Ocean coasts, see Fig. B.8, with Norway and Chile the two countries having the most abundant natural resources. For example, the coast along Norway possesses 50–60 million tons of wide seaweeds with 7 million tons of seaweed washed to the shore each year. Chile has a coastline of more than 4800 km where the cool and clean Pacific Ocean is ideal for brown seaweed to grow. Food and Agricultural Organization (FAO) estimates that Chile harvests more than 500 000 tons of wild seaweeds each year and nowadays, the majority of them are exported to China where alginate is extracted from them in companies such as Qingdao Brightmoon Seaweed Group, which is the largest seaweed processing company in the world. Once alginate is extracted from the seaweed biomass, it is dissolved in water and extruded in the wet spinning process to form fibers, which are further processed into nonwoven fabrics and wound dressings that can be used on chronic ulcerative wounds all over the world, fulfilling the promises of "utilizing marine resources, benefiting human health".

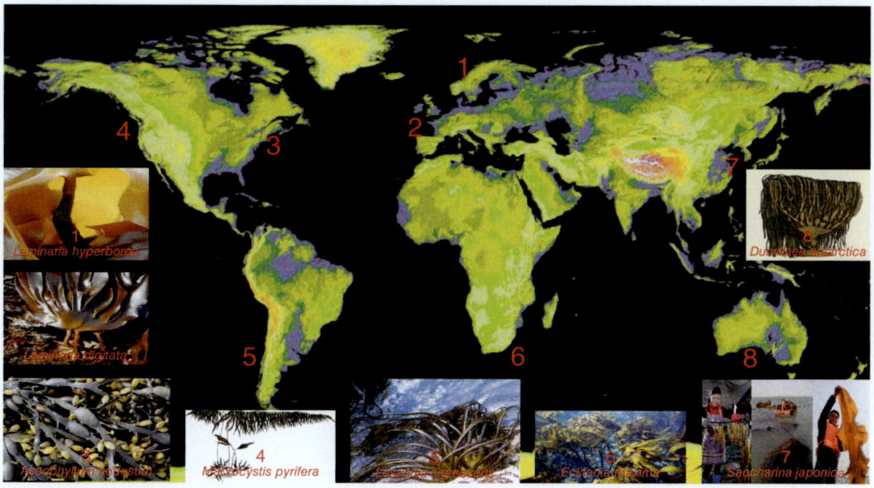

Figure B.8 Distribution of brown seaweeds around the world.

Index

Alginate Fibers and Wound Dressings: Seaweed Derived Natural Therapy, First Edition. Yimin Qin.